Succeed

Eureka Math®
Grade 4
Modules 5–7

Published by Great Minds®.

Copyright © 2018 Great Minds®.

Printed in the U.S.A.
This book may be purchased from the publisher at eureka-math.org.
2 3 4 5 6 7 8 9 10 CCR 24 23 22 21

ISBN 978-1-64054-091-0

G4-M5-M7-S-06.2018

Learn ♦ Practice ♦ Succeed

Eureka Math® student materials for *A Story of Units*® (K–5) are available in the *Learn, Practice, Succeed* trio. This series supports differentiation and remediation while keeping student materials organized and accessible. Educators will find that the *Learn, Practice,* and *Succeed* series also offers coherent—and therefore, more effective—resources for Response to Intervention (RTI), extra practice, and summer learning.

Learn

Eureka Math Learn serves as a student's in-class companion where they show their thinking, share what they know, and watch their knowledge build every day. *Learn* assembles the daily classwork—Application Problems, Exit Tickets, Problem Sets, templates—in an easily stored and navigated volume.

Practice

Each *Eureka Math* lesson begins with a series of energetic, joyous fluency activities, including those found in *Eureka Math Practice.* Students who are fluent in their math facts can master more material more deeply. With *Practice,* students build competence in newly acquired skills and reinforce previous learning in preparation for the next lesson.

Together, *Learn* and *Practice* provide all the print materials students will use for their core math instruction.

Succeed

Eureka Math Succeed enables students to work individually toward mastery. These additional problem sets align lesson by lesson with classroom instruction, making them ideal for use as homework or extra practice. Each problem set is accompanied by a Homework Helper, a set of worked examples that illustrate how to solve similar problems.

Teachers and tutors can use *Succeed* books from prior grade levels as curriculum-consistent tools for filling gaps in foundational knowledge. Students will thrive and progress more quickly as familiar models facilitate connections to their current grade-level content.

Students, families, and educators:

Thank you for being part of the *Eureka Math*® community, where we celebrate the joy, wonder, and thrill of mathematics.

Nothing beats the satisfaction of success—the more competent students become, the greater their motivation and engagement. The *Eureka Math Succeed* book provides the guidance and extra practice students need to shore up foundational knowledge and build mastery with new material.

What is in the Succeed *book?*

Eureka Math Succeed books deliver supported practice sets that parallel the lessons of *A Story of Units*®. Each *Succeed* lesson begins with a set of worked examples, called *Homework Helpers*, that illustrate the modeling and reasoning the curriculum uses to build understanding. Next, students receive scaffolded practice through a series of problems carefully sequenced to begin from a place of confidence and add incremental complexity.

How should Succeed *be used?*

The collection of *Succeed* books can be used as differentiated instruction, practice, homework, or intervention. When coupled with *Affirm*®, *Eureka Math*'s digital assessment system, *Succeed* lessons enable educators to give targeted practice and to assess student progress. *Succeed*'s perfect alignment with the mathematical models and language used across *A Story of Units* ensures that students feel the connections and relevance to their daily instruction, whether they are working on foundational skills or getting extra practice on the current topic.

Where can I learn more about Eureka Math *resources?*

The Great Minds® team is committed to supporting students, families, and educators with an ever-growing library of resources, available at eureka-math.org. The website also offers inspiring stories of success in the *Eureka Math* community. Share your insights and accomplishments with fellow users by becoming a *Eureka Math* Champion.

Best wishes for a year filled with Eureka moments!

Jill Diniz

Jill Diniz
Director of Mathematics
Great Minds

Contents

Module 5: Fraction Equivalence, Ordering, and Operations

Module 6: Decimal Fractions

Module 7: Exploring Measurement with Multiplication

Topic C: Investigation of Measurements Expressed as Mixed Numbers

Topic D: Year in Review

Grade 4
Module 5

1. Draw a number bond, and write the number sentence to match each tape diagram.

   a.

   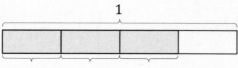

   $$\frac{3}{4} = \frac{1}{4} + \frac{1}{4} + \frac{1}{4}$$

   The rectangle represents 1 and is partitioned into 4 equal units. Each unit is equal to 1 fourth.

   I can decompose any fraction into unit fractions. 3 fourths is composed of 3 units of 1 fourth.

   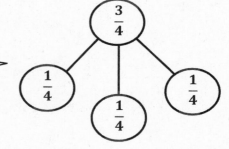

   b.

   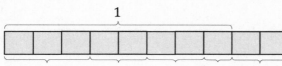

   $$\frac{10}{8} = \frac{3}{8} + \frac{2}{8} + \frac{2}{8} + \frac{1}{8} + \frac{2}{8}$$

   I can rename a fraction greater than 1, such as $\frac{10}{8}$, as a whole number and a fraction, $1\frac{2}{8}$.

   I know the fractional unit is eighths. I count 8 equal units bracketed as 1 whole.

   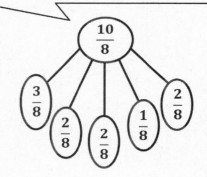

2. Draw and label tape diagrams to match each number sentence.

   a. $\frac{11}{6} = \frac{3}{6} + \frac{2}{6} + \frac{2}{6} + \frac{4}{6}$

   b. $1\frac{2}{12} = \frac{7}{12} + \frac{4}{12} + \frac{3}{12}$

   I know the unit is twelfths. I partition my tape diagram into 12 equal units to represent the whole. I draw 2 more twelfths.

   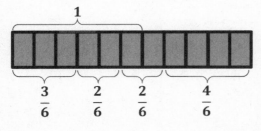

   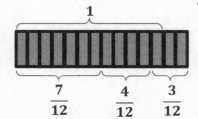

1. Decompose each fraction modeled by a tape diagram as a sum of unit fractions. Write the equivalent multiplication sentence.

   a.

   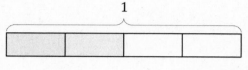

   $$\frac{2}{4} = \frac{1}{4} + \frac{1}{4} \qquad \frac{2}{4} = 2 \times \frac{1}{4}$$

   > There are 2 copies of $\frac{1}{4}$ shaded, so I write $2 \times \frac{1}{4}$.

   > I can multiply fourths like I multiply any other unit. 1 banana times 2 is 2 bananas and 1 ten times 2 is 2 tens, so 1 fourth times 2 is 2 fourths.

   b.

   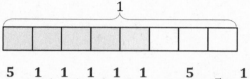

   $$\frac{5}{8} = \frac{1}{8} + \frac{1}{8} + \frac{1}{8} + \frac{1}{8} + \frac{1}{8} \qquad \frac{5}{8} = 5 \times \frac{1}{8}$$

   > I can add 1 eighth 5 times. Whew! That's a lot of writing! Or I can multiply to show 5 copies of $\frac{1}{8}$.

2. The tape diagram models a fraction greater than 1. Write the fraction greater than 1 as the sum of two products.

   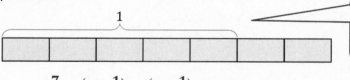

   > This bracket identifies the whole. This tape diagram models a fraction greater than 1.

   $$\frac{7}{5} = \left(5 \times \frac{1}{5}\right) + \left(2 \times \frac{1}{5}\right)$$

   > I see in the tape diagram that $\frac{7}{5}$ is the same as $1\frac{2}{5}$. I can use the distributive property to express the whole part and the fractional part as 2 different multiplication expressions.

3. Draw a tape diagram to model $\frac{9}{8}$. Record the decomposition of $\frac{9}{8}$ into unit fractions as a multiplication sentence.

   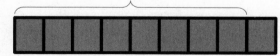

   $$\frac{9}{8} = 9 \times \frac{1}{8}$$

Lesson 3:   Decompose non-unit fractions and represent them as a whole number times a unit fraction using tape diagrams.

© 2018 Great Minds®. eureka-math.org

11

1. The total length of each tape diagram represents 1. Decompose the shaded unit fractions as the sum of smaller unit fractions in at least two different ways.

a.

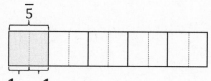

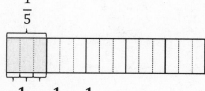

$$\frac{1}{5} = \frac{1}{10} + \frac{1}{10}$$

$$\frac{1}{15} + \frac{1}{15} + \frac{1}{15} = \frac{1}{5}$$

> After decomposing each fifth into 2 equal parts, the new unit is tenths.

b.

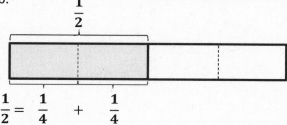

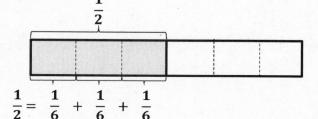

$$\frac{1}{2} = \frac{1}{4} + \frac{1}{4}$$

$$\frac{1}{2} = \frac{1}{6} + \frac{1}{6} + \frac{1}{6}$$

2. Draw a tape diagram to prove $\frac{2}{3} = \frac{4}{6}$.

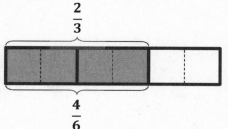

> I know that $\frac{2}{3}$ and $\frac{4}{6}$ are equal because they take up the same amount of space.

3. Show that $\frac{1}{2}$ is equivalent to $\frac{4}{8}$ using a tape diagram and a number sentence.

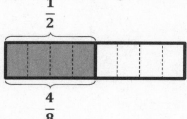

$$\frac{1}{2} = 4 \times \frac{1}{8}$$

> I quadrupled the number of units within each half, which I can record as a multiplication sentence.

EUREKA MATH

Lesson 4: Decompose fractions into sums of smaller unit fractions using tape diagrams.

15

© 2018 Great Minds®. eureka-math.org

1. Draw horizontal line(s) to decompose the rectangle into 2 rows. Use the model to name the shaded area as both a sum of unit fractions and as a multiplication sentence.

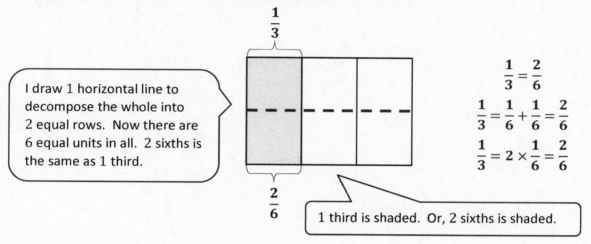

I draw 1 horizontal line to decompose the whole into 2 equal rows. Now there are 6 equal units in all. 2 sixths is the same as 1 third.

$$\frac{1}{3} = \frac{2}{6}$$

$$\frac{1}{3} = \frac{1}{6} + \frac{1}{6} = \frac{2}{6}$$

$$\frac{1}{3} = 2 \times \frac{1}{6} = \frac{2}{6}$$

1 third is shaded. Or, 2 sixths is shaded.

2. Draw area models to show the decompositions represented by the number sentences below. Represent the decomposition as a sum of unit fractions and as a multiplication sentence.

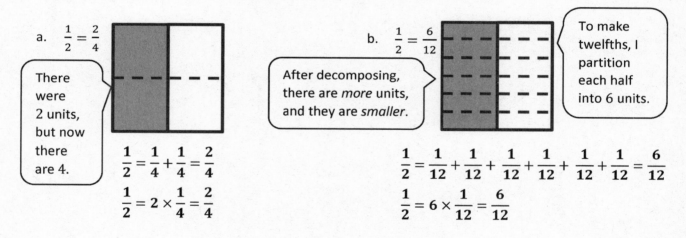

a. $\frac{1}{2} = \frac{2}{4}$

There were 2 units, but now there are 4.

$$\frac{1}{2} = \frac{1}{4} + \frac{1}{4} = \frac{2}{4}$$

$$\frac{1}{2} = 2 \times \frac{1}{4} = \frac{2}{4}$$

b. $\frac{1}{2} = \frac{6}{12}$

After decomposing, there are *more* units, and they are *smaller*.

To make twelfths, I partition each half into 6 units.

$$\frac{1}{2} = \frac{1}{12} + \frac{1}{12} + \frac{1}{12} + \frac{1}{12} + \frac{1}{12} + \frac{1}{12} = \frac{6}{12}$$

$$\frac{1}{2} = 6 \times \frac{1}{12} = \frac{6}{12}$$

3. Explain why $\frac{1}{12} + \frac{1}{12} + \frac{1}{12} + \frac{1}{12} + \frac{1}{12} + \frac{1}{12}$ is the same as $\frac{1}{2}$.

    Sample Student Response:

    I see in the area model that I drew that 6 twelfths takes up the same space as 1 half. 6 twelfths and 1 half have exactly the same area.

EUREKA MATH   **Lesson 5:**   Decompose unit fractions using area models to show equivalence.   21

1. The rectangle represents 1. Draw horizontal line(s) to decompose the rectangle into *twelfths*. Use the model to name the shaded area as a sum and as a product of unit fractions. Use parentheses to show the relationship between the number sentences.

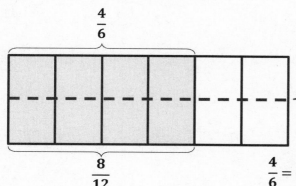

$$\frac{4}{6}$$

$$\frac{8}{12}$$

$$\frac{4}{6} = \frac{8}{12}$$

4 sixths are shaded. I draw one line to partition sixths into twelfths. 8 twelfths are shaded.

I write addition and multiplication sentences using unit fractions.

$$\frac{1}{6} + \frac{1}{6} + \frac{1}{6} + \frac{1}{6} = \left(\frac{1}{12} + \frac{1}{12}\right) + \left(\frac{1}{12} + \frac{1}{12}\right) + \left(\frac{1}{12} + \frac{1}{12}\right) + \left(\frac{1}{12} + \frac{1}{12}\right) = \frac{8}{12}$$

$$\left(\frac{1}{12} + \frac{1}{12}\right) + \left(\frac{1}{12} + \frac{1}{12}\right) + \left(\frac{1}{12} + \frac{1}{12}\right) + \left(\frac{1}{12} + \frac{1}{12}\right) = \left(2 \times \frac{1}{12}\right) + \left(2 \times \frac{1}{12}\right) + \left(2 \times \frac{1}{12}\right) + \left(2 \times \frac{1}{12}\right) = \frac{8}{12}$$

$$\frac{4}{6} = 8 \times \frac{1}{12} = \frac{8}{12}$$

2. Draw an area model to show the decompositions represented by $\frac{2}{3} = \frac{6}{9}$. Express $\frac{2}{3} = \frac{6}{9}$ as a sum and product of unit fractions. Use parentheses to show the relationship between the number sentences.

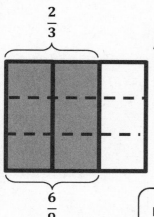

$$\frac{2}{3}$$

$$\frac{6}{9}$$

I draw thirds vertically and partition the thirds into ninths with two horizontal lines.

$$\frac{2}{3} = \frac{6}{9}$$

$$\frac{1}{3} + \frac{1}{3} = \left(\frac{1}{9} + \frac{1}{9} + \frac{1}{9}\right) + \left(\frac{1}{9} + \frac{1}{9} + \frac{1}{9}\right) = \frac{6}{9}$$

$$\left(\frac{1}{9} + \frac{1}{9} + \frac{1}{9}\right) + \left(\frac{1}{9} + \frac{1}{9} + \frac{1}{9}\right) = \left(3 \times \frac{1}{9}\right) + \left(3 \times \frac{1}{9}\right) = \frac{6}{9}$$

I write parentheses that show the decomposition of $\frac{1}{3}$. Just as the area model shows 1 third partitioned into 3 ninths, so do the parentheses.

Lesson 6: Decompose fractions using area models to show equivalence.

25

EUREKA MATH®

Name _____   Date _____

1. Each rectangle represents 1. Draw horizontal lines to decompose each rectangle into the fractional units as indicated. Use the model to give the shaded area as a sum and as a product of unit fractions. Use parentheses to show the relationship between the number sentences. The first one has been partially done for you.

a. Tenths

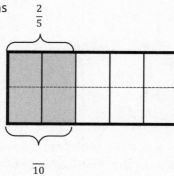

$$\frac{2}{5} = \frac{4}{}$$

$$\frac{}{5} + \frac{}{5} = \left(\frac{1}{10} + \frac{1}{10}\right) + \left(\frac{1}{10} + \frac{1}{10}\right) = \frac{4}{}$$

$$\left(\frac{1}{10} + \frac{1}{10}\right) + \left(\frac{1}{10} + \frac{1}{10}\right) = \left(2 \times \frac{}{}\right) + \left(2 \times \frac{}{}\right) = \frac{4}{}$$

$$\frac{2}{5} = 4 \times \frac{}{} = \frac{4}{}$$

b. Eighths

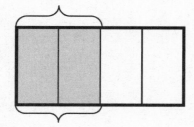

EUREKA MATH®

c. Fifteenths

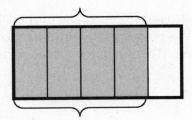

2. Draw area models to show the decompositions represented by the number sentences below. Express each as a sum and product of unit fractions. Use parentheses to show the relationship between the number sentences.

a. $\frac{2}{3} = \frac{4}{6}$

b. $\frac{4}{5} = \frac{8}{10}$

Lesson 6:   Decompose fractions using area models to show equivalence.

EUREKA MATH®

3.   Step 1:   Draw an area model for a fraction with units of thirds, fourths, or fifths.

       Step 2:   Shade in more than one fractional unit.

       Step 3:   Partition the area model again to find an equivalent fraction.

       Step 4:   Write the equivalent fractions as a number sentence. (If you have written a number sentence like this one already in this Homework, start over.)

Each rectangle represents 1.

1. The shaded unit fractions have been decomposed into smaller units. Express the equivalent fractions in a number sentence using multiplication.

a.

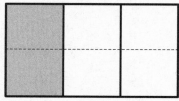

$$\frac{1}{3} = \frac{1 \times 2}{3 \times 2} = \frac{2}{6}$$

b.

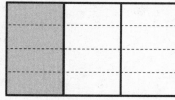

$$\frac{1}{3} = \frac{1 \times 4}{3 \times 4} = \frac{4}{12}$$

> The numerator is 1.
> The denominator is 3.

> I can multiply the numerator (number of fractional units selected) and the denominator (the fractional unit) by 4 to make an equivalent fraction.

2. Decompose the shaded fraction into smaller units using the area model. Express the equivalent fractions in a number sentence using multiplication.

> The area model shows that $\frac{1}{6}$ equals $\frac{3}{18}$.

> As I multiply, the size of the units gets smaller.

$$\frac{1}{6} = \frac{1 \times 3}{6 \times 3} = \frac{3}{18}$$

3. Draw three different area models to represent 1 half by shading.

Decompose the shaded fraction into (a) fourths, (b) sixths, and (c) eighths.

Use multiplication to show how each fraction is equivalent to 1 half.

a.

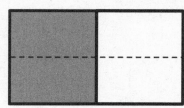

$$\frac{1}{2} = \frac{1 \times 2}{2 \times 2} = \frac{2}{4}$$

> The number of units doubled.

b.

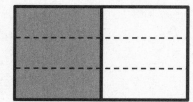

$$\frac{1}{2} = \frac{1 \times 3}{2 \times 3} = \frac{3}{6}$$

> The number of units tripled.

c.

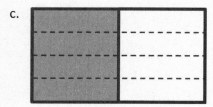

$$\frac{1}{2} = \frac{1 \times 4}{2 \times 4} = \frac{4}{8}$$

> The number of units quadrupled.

Each rectangle represents 1.

1. Compose the shaded fraction into larger fractional units. Express the equivalent fractions in a number sentence using division.

a.

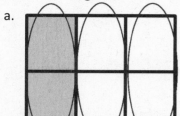

b.

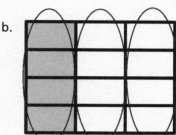

> 2 units are shaded. I make groups of 2. Sixths are composed as thirds.

$$\frac{2}{6} = \frac{2 \div 2}{6 \div 2} = \frac{1}{3}$$

> I divide the numerator and denominator by 2.

$$\frac{4}{12} = \frac{4 \div 4}{12 \div 4} = \frac{1}{3}$$

> When I compose thirds, the number of units decreases. I make a larger unit.

2.

a. In the first model, show 2 tenths. In the second area model, show 3 fifteenths. Show how both fractions can be composed, or renamed, as the same unit fraction.

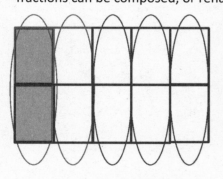   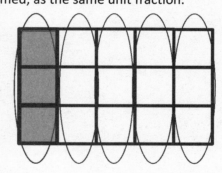

> Before I draw my model, I identify the larger unit fraction. I know 3 fifteenths is the same as $\frac{1 \times 3}{5 \times 3}$.

2 *tenths* = 1 *fifth*   3 *fifteenths* = 1 *fifth*

b. Express the equivalent fractions in a number sentence using division.

___ ___ ___   ___ ___ ___

Each rectangle represents 1.

1. Compose the shaded fraction into larger fractional units. Express the equivalent fractions in a number sentence using division.

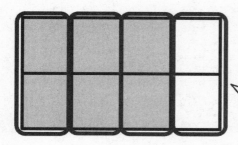

$$\frac{6}{8} = \frac{6 \div 2}{8 \div 2} = \frac{3}{4}$$

This work is a lot like what I did in Lesson 9. However, once I compose units, the renamed fraction is not a unit fraction.

2. Draw an area model to represent the number sentence below.

$$\frac{4}{14} = \frac{4 \div 2}{14 \div 2} = \frac{2}{7}$$

Looking at the numerator and denominator, I draw 14 units and shade 4 units.

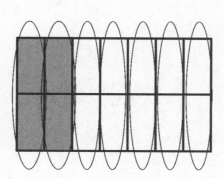

Looking at the divisor, $\frac{2}{2}$, I circle groups of 2. I make 7 groups. 2 sevenths are shaded.

3. Use division to rename the fraction below. Draw a model if that helps you. See if you can use the largest common factor.

$$\frac{8}{20} = \frac{8 \div 4}{20 \div 4} = \frac{2}{5}$$

I could choose 2, but the largest common factor is 4.

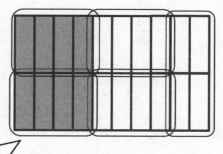

Whether I compose units vertically or horizontally, I get the same answer!

Name _____   Date _____

1. Label each number line with the fractions shown on the tape diagram. Circle the fraction that labels the point on the number line that also names the shaded part of the tape diagram.

a.

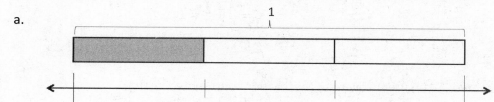

b.

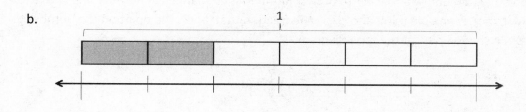

c.

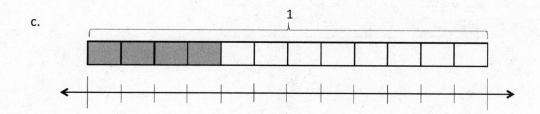

EUREKA MATH

Lesson 11:   Explain fraction equivalence using a tape diagram and the number line, and relate that to the use of multiplication and division.

© 2018 Great Minds®. eureka-math.org

53

2. Write number sentences using multiplication to show:

   a. The fraction represented in 1(a) is equivalent to the fraction represented in 1(b).

   b. The fraction represented in 1(a) is equivalent to the fraction represented in 1(c).

3. Use each shaded tape diagram below as a ruler to draw a number line. Mark each number line with the fractional units shown on the tape diagram, and circle the fraction that labels the point on the number line that also names the shaded part of the tape diagram.

   a.

   1

   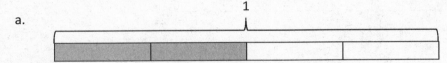

   b.

   1

   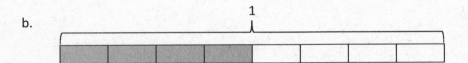

   c.

   1

   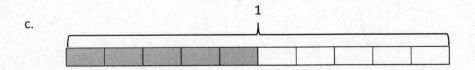

Lesson 11: Explain fraction equivalence using a tape diagram and the number line, and relate that to the use of multiplication and division.

EUREKA MATH

4. Write a number sentence using division to show the fraction represented in 3(a) is equivalent to the fraction represented in 3(b).

5. a. Partition a number line from 0 to 1 into fourths. Decompose $\frac{3}{4}$ into 6 equal lengths.

   b. Write a number sentence using multiplication to show what fraction represented on the number line is equivalent to $\frac{3}{4}$.

   c. Write a number sentence using division to show what fraction represented on the number line is equivalent to $\frac{3}{4}$.

Lesson 11:   Explain fraction equivalence using a tape diagram and the number line, and relate that to the use of multiplication and division.

© 2018 Great Minds®. eureka-math.org

55

1.

a. Plot the following points on the number line without measuring.

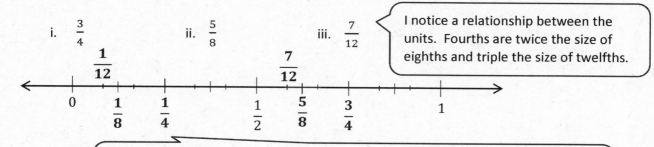

i. $\frac{3}{4}$ ii. $\frac{5}{8}$ iii. $\frac{7}{12}$

> I notice a relationship between the units. Fourths are twice the size of eighths and triple the size of twelfths.

> I use benchmark fractions I know to plot twelfths. After marking fourths, I know that 1 fourth is the same as 3 twelfths, so I decompose each fourth into 3 units to make twelfths.

b. Use the number line in part (a) to compare the fractions by writing >, <, or = on the lines.

i. $\frac{3}{4}$ ___>___ $\frac{1}{2}$ ii. $\frac{7}{12}$ ___<___ $\frac{5}{8}$

c. Explain how you plotted the points in Part (a).

Sample Student Response:

The number line was partitioned into halves. I doubled the units to make fourths. I plotted 3 fourths. I doubled the units again to make eighths. Knowing that 1 half and 4 eighths are equivalent fractions, I simply counted on 1 more eighth to plot 5 eighths. Lastly, I thought about twelfths and fourths. 1 fourth is the same as 3 twelfths. I marked twelfths by partitioning each fourth into 3 units. I plotted 7 twelfths.

2. Compare the fractions given below by writing < or > on the line.

Give a brief explanation for each answer referring to the benchmarks of $0, \frac{1}{2},$ and/or 1.

$\frac{5}{8}$ ___>___ $\frac{6}{10}$

Possible student response:

If I think about eighths, I know that 1 half is equal to 4 eighths. Therefore, 5 eighths is 1 eighth greater than 1 half.

I also know that 5 tenths is equal to 1 half. 6 tenths is 1 tenth greater than 1 half. Comparing the size of the units, I know that 1 eighth is more than 1 tenth. So, 5 eighths is greater than 6 tenths.

EUREKA MATH®

Lesson 12:   Reason using benchmarks to compare two fractions on the number line.

© 2018 Great Minds®. eureka-math.org

57

Name _____   Date _____

1. a.   Plot the following points on the number line without measuring.

i. $\frac{2}{3}$ ii. $\frac{1}{6}$ iii. $\frac{4}{10}$

0                                    $\frac{1}{2}$ 1

b.   Use the number line in Part (a) to compare the fractions by writing >, <, or = on the lines.

i. $\frac{2}{3}$ _____ $\frac{1}{2}$ ii. $\frac{4}{10}$ _____ $\frac{1}{6}$

2. a.   Plot the following points on the number line without measuring.

i. $\frac{5}{12}$ ii. $\frac{3}{4}$ iii. $\frac{2}{6}$

0                                    $\frac{1}{2}$ 1

b.   Select two fractions from Part (a), and use the given number line to compare them by writing >, <, or =.

c.   Explain how you plotted the points in Part (a).

EUREKA
MATH®

3. Compare the fractions given below by writing > or < on the lines.

   Give a brief explanation for each answer referring to the benchmark of 0, $\frac{1}{2}$, and 1.

a. $\frac{1}{2}$ ——————— $\frac{1}{4}$

b. $\frac{6}{8}$ ——————— $\frac{1}{2}$

c. $\frac{3}{4}$ ——————— $\frac{3}{5}$

d. $\frac{4}{6}$ ——————— $\frac{9}{12}$

e. $\frac{2}{3}$ ——————— $\frac{1}{4}$

f. $\frac{4}{5}$ ——————— $\frac{8}{12}$

g. $\frac{1}{3}$ ——————— $\frac{3}{6}$

h. $\frac{7}{8}$ ——————— $\frac{3}{5}$

i. $\frac{51}{100}$ ——————— $\frac{5}{10}$

j. $\frac{8}{14}$ ——————— $\frac{49}{100}$

Lesson 12:   Reason using benchmarks to compare two fractions on the number line.

EUREKA MATH

1. Place the following fractions on the number line given.

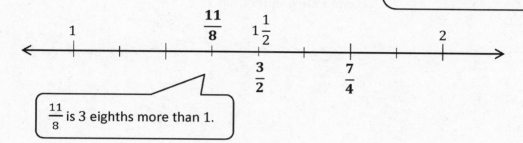

$\frac{8}{4}$ is equal to 2. Therefore, $\frac{7}{4}$ is 1 fourth less than 2.

a. $\frac{7}{4}$ b. $\frac{3}{2}$ c. $\frac{11}{8}$

I can draw a number bond, breaking $\frac{11}{8}$ into $\frac{8}{8}$ and $\frac{3}{8}$.

$\frac{11}{8}$ is 3 eighths more than 1.

2. Use the number line in Problem 1 to compare the fractions by writing <, >, or = on the lines.

 a. $1\frac{3}{4}$ __>__ $1\frac{1}{2}$ b. $1\frac{3}{8}$ __<__ $1\frac{3}{4}$

Using the benchmark $\frac{1}{2}$, I compare the fractions. $1\frac{3}{8}$ is less than 1 and 1 half, while $1\frac{3}{4}$ is more than 1 and 1 half.

3. Use the number line in Problem 1 to explain the reasoning you used when determining whether $\frac{11}{8}$ or $\frac{7}{4}$ was greater.

 Sample Student Response:

 After I plotted $\frac{11}{8}$ and $\frac{7}{4}$, I noticed that $\frac{7}{4}$ was greater than $1\frac{1}{2}$, whereas $\frac{11}{8}$ is less than $1\frac{1}{2}$.

EUREKA MATH Lesson 13: Reason using benchmarks to compare two fractions on the number line.   61

© 2018 Great Minds®. eureka-math.org

4. Compare the fractions given below by writing < or > on the lines. Give a brief explanation for each answer referring to benchmarks.

a. $\dfrac{5}{4}$ ___>___ $\dfrac{9}{10}$

b. $\dfrac{7}{12}$ ___<___ $\dfrac{7}{6}$ ── I use two different benchmarks to compare these fractions.

$\dfrac{5}{4}$ *is greater than* **1**.

$\dfrac{9}{10}$ *is less than* **1**.

$\dfrac{7}{12}$ *is one twelfth greater than* $\dfrac{1}{2}$.

$\dfrac{7}{6}$ *is one sixth greater than* **1**.

Lesson 13: Reason using benchmarks to compare two fractions on the number line.

EUREKA MATH

Name _____ Date _____

1. Place the following fractions on the number line given.

   a. $\frac{3}{2}$

   b. $\frac{9}{5}$

   c. $\frac{14}{10}$

2. Use the number line in Problem 1 to compare the fractions by writing >, <, or = on the lines.

   a. $1\frac{1}{6}$ _____ $1\frac{4}{12}$

   b. $1\frac{1}{2}$ _____ $1\frac{4}{5}$

3. Place the following fractions on the number line given.

   a. $\frac{12}{9}$

   b. $\frac{6}{5}$

   c. $\frac{18}{15}$

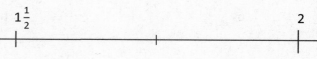

4. Use the number line in Problem 3 to explain the reasoning you used when determining whether $\frac{12}{9}$ or $\frac{18}{15}$ was greater.

EUREKA MATH

Lesson 13: Reason using benchmarks to compare two fractions on the number line.

© 2018 Great Minds®. eureka-math.org

63

5. Compare the fractions given below by writing > or < on the lines. Give a brief explanation for each answer referring to benchmarks.

a. $\frac{2}{5}$ —————— $\frac{6}{8}$

b. $\frac{6}{10}$ —————— $\frac{5}{6}$

c. $\frac{6}{4}$ —————— $\frac{7}{8}$

d. $\frac{1}{4}$ —————— $\frac{8}{12}$

e. $\frac{14}{12}$ —————— $\frac{11}{6}$

f. $\frac{8}{9}$ —————— $\frac{3}{2}$

g. $\frac{7}{8}$ —————— $\frac{11}{10}$

h. $\frac{3}{4}$ —————— $\frac{4}{3}$

i. $\frac{3}{8}$ —————— $\frac{3}{2}$

j. $\frac{9}{6}$ —————— $\frac{16}{12}$

EUREKA MATH

1. Compare the pairs of fractions by reasoning about the size of the units. Use >, <, or =.

   a.   1 fourth   __>__   1 eighth   b.   2 thirds   __>__   2 fifths

   > I envision a tape diagram. 1 fourth is double the size of 1 eighth.

   > When I'm comparing the same number of units, I consider the size of the fractional unit. Thirds are bigger than fifths.

2. Compare by reasoning about the following pair of fractions with related numerators. Use >, <, or =. Explain your thinking using words, pictures, or numbers.

   $\dfrac{3}{7}$ __>__ $\dfrac{6}{15}$

   > To compare, I can make the numerators the same.

   3 *sevenths are equal to 6 fourteenths. Fourteenths are greater than fifteenths. So, 3 sevenths are greater than 6 fifteenths.*

3. Draw two tape diagrams to model and compare $1\dfrac{3}{4}$ and $1\dfrac{8}{12}$.

   $1\dfrac{3}{4}$ __>__ $1\dfrac{8}{12}$

   > The model shows that $\dfrac{9}{12}$ is equal to $\dfrac{3}{4}$. So, $\dfrac{8}{12}$ is less.

   > I'm careful to make each tape diagram the same size.

   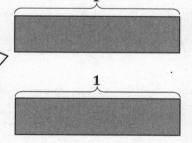
   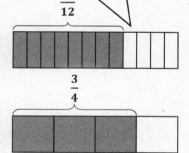

4. Draw one number line to model the pair of fractions with related denominators. Use >, <, or = to compare.

   $\dfrac{3}{12}$ __<__ $\dfrac{2}{6}$

   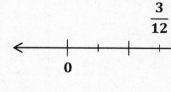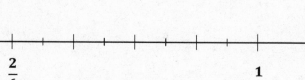

Name _____ Date _____

1. Compare the pairs of fractions by reasoning about the size of the units. Use >, <, or =.

   a. 1 third _____ 1 sixth

   b. 2 halves _____ 2 thirds

   c. 2 fourths _____ 2 sixths

   d. 5 eighths _____ 5 tenths

2. Compare by reasoning about the following pairs of fractions with the same or related numerators. Use >, <, or =. Explain your thinking using words, pictures, or numbers. Problem 2(b) has been done for you.

   a. $\frac{3}{6}$ _____ $\frac{3}{7}$

   b. $\frac{2}{5} < \frac{4}{9}$

   because $\frac{2}{5} = \frac{4}{10}$

   4 tenths is less

   than 4 ninths because

   tenths are smaller than ninths.

   $\frac{2}{5} = \frac{4}{10}$

   $\frac{4}{9}$

   c. $\frac{3}{11}$ _____ $\frac{3}{13}$

   d. $\frac{5}{7}$ _____ $\frac{10}{13}$

3. Draw two tape diagrams to model each pair of the following fractions with related denominators. Use >, <, or = to compare.

    a. $\frac{3}{4}$ _____ $\frac{7}{12}$

    b. $\frac{2}{4}$ _____ $\frac{1}{8}$

    c. $1\frac{4}{10}$ _____ $1\frac{3}{5}$

EUREKA MATH

4. Draw one number line to model each pair of fractions with related denominators. Use >, <, or = to compare.

   a. $\frac{3}{4}$ _____ $\frac{5}{8}$

   b. $\frac{11}{12}$ _____ $\frac{3}{4}$

   c. $\frac{4}{5}$ _____ $\frac{7}{10}$

   d. $\frac{8}{9}$ _____ $\frac{2}{3}$

5. Compare each pair of fractions using >, <, or =. Draw a model if you choose to.

   a. $\frac{1}{7}$ _____ $\frac{2}{7}$

   b. $\frac{5}{7}$ _____ $\frac{11}{14}$

   c. $\frac{7}{10}$ _____ $\frac{3}{5}$

   d. $\frac{2}{3}$ _____ $\frac{9}{15}$

   e. $\frac{3}{4}$ _____ $\frac{9}{12}$

   f. $\frac{5}{3}$ _____ $\frac{5}{2}$

Lesson 14: Find common units or number of units to compare two fractions.

69

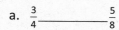

EUREKA MATH

6. Simon claims $\frac{4}{9}$ is greater than $\frac{1}{3}$. Ted thinks $\frac{4}{9}$ is less than $\frac{1}{3}$. Who is correct? Support your answer with a picture.

Lesson 14:   Find common units or number of units to compare two fractions.

EUREKA
MATH

1. Draw an area model for the pair of fractions, and use it to compare the two fractions by writing <, >, or = on the line.

$$\frac{4}{5} \; \underline{<} \; \frac{6}{7}$$

$$\frac{28}{35} \; \underline{<} \; \frac{30}{35}$$

$$\frac{4 \times 7}{5 \times 7} = \frac{28}{35}$$

> I use two area models that are exactly the same size to find like units. After partitioning, I have 35 units in each model. Now I can compare!

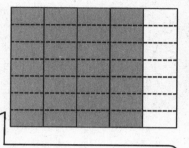

$$\frac{6 \times 5}{7 \times 5} = \frac{30}{35}$$

> I represent fifths with vertical lines and then partition fifths by drawing horizontal lines.

> I represent sevenths with horizontal lines and then partition sevenths by drawing vertical lines.

2. Rename the fractions below using multiplication, and then compare by writing <, >, or =.

$$\frac{5}{8} \; \underline{<} \; \frac{9}{12} \qquad \frac{5 \times 12}{8 \times 12} = \frac{60}{96} \qquad \frac{9 \times 8}{12 \times 8} = \frac{72}{96}$$

> Whew! That would have been a lot of units to draw in an area model!

$$\frac{60}{96} \; \underline{<} \; \frac{72}{96}$$

> Using multiplication to make common units is quick and precise. It is best to compare fractions when the units are the same.

3. Use any method to compare the fractions below. Record your answer using $<$, $>$, or $=$.

$$\frac{5}{3} \underline{\ <\ } \frac{9}{5}$$

$$\frac{3}{3} = \frac{5}{5}$$

$$\frac{2}{3} < \frac{4}{5}$$

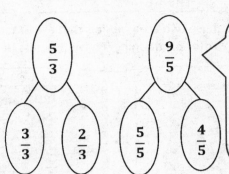

I use number bonds to decompose fractions greater than 1. This lets me focus on the fractional parts, $\frac{2}{3}$ and $\frac{4}{5}$, to compare since $\frac{3}{3}$ and $\frac{5}{5}$ are equivalent.

I use benchmarks to compare. $\frac{4}{5}$ is closer to 1 than $\frac{2}{3}$ because fifths are smaller than thirds.

Lesson 15:   Find common units or number of units to compare two fractions.   EUREKA MATH

Name _____   Date _____

1. Draw an area model for each pair of fractions, and use it to compare the two fractions by writing
    >, <, or = on the line. The first two have been partially done for you. Each rectangle represents 1.

a. $\frac{1}{2}$ ___ < ___ $\frac{3}{5}$

$\frac{1 \times 5}{2 \times 5} = \frac{5}{10}$ $\frac{3 \times 2}{5 \times 2} = \frac{6}{10}$

$\frac{5}{10} < \frac{6}{10}$, so $\frac{1}{2} < \frac{3}{5}$

b. $\frac{2}{3}$ _____ $\frac{3}{4}$

c. $\frac{4}{6}$ _____ $\frac{5}{8}$

d. $\frac{2}{7}$ _____ $\frac{3}{5}$

e. $\frac{4}{6}$ _____ $\frac{6}{9}$

f. $\frac{4}{5}$ _____ $\frac{5}{6}$

EUREKA
MATH®

2. Rename the fractions, as needed, using multiplication in order to compare each pair of fractions by writing >, <, or =.

   a. $\frac{2}{3}$ _____ $\frac{2}{4}$

   b. $\frac{4}{7}$ _____ $\frac{1}{2}$

   c. $\frac{5}{4}$ _____ $\frac{9}{8}$

   d. $\frac{8}{12}$ _____ $\frac{5}{8}$

3. Use any method to compare the fractions. Record your answer using >, <, or =.

   a. $\frac{8}{9}$ _____ $\frac{2}{3}$

   b. $\frac{4}{7}$ _____ $\frac{4}{5}$

   c. $\frac{3}{2}$ _____ $\frac{9}{6}$

   d. $\frac{11}{7}$ _____ $\frac{5}{3}$

EUREKA MATH

4. Explain which method you prefer using to compare fractions. Provide an example using words, pictures, or numbers.

Lesson 15:   Find common units or number of units to compare two fractions.

© 2018 Great Minds®. eureka-math.org

75

Solve.

1. 5 sixths − 3 sixths = **2 sixths**

> The units in both numbers are the same, so I can think "5 − 3 = 2," so 5 sixths − 3 sixths = 2 sixths.

> I can rewrite the number sentence using fractions.
> $$\frac{5}{6} - \frac{3}{6} = \frac{2}{6}$$

2. 1 sixth + 4 sixths = **5 sixths**

> If I know that 1 + 4 = 5, then 1 sixth + 4 sixths = 5 sixths.

Solve. Use a number bond to rename the sum or difference as a mixed number. Then, draw a number line to model your answer.

3. $\frac{12}{6} - \frac{5}{6} = \frac{7}{6} = 1\frac{1}{6}$

    $\frac{6}{6}$ $\frac{1}{6}$

> I can rename $\frac{7}{6}$ as a mixed number using a number bond to separate, or decompose, $\frac{7}{6}$ into a whole number and a fraction. $\frac{6}{6}$ is the whole, and the fractional part is $\frac{1}{6}$.

4. $\frac{5}{6} + \frac{5}{6} = \frac{10}{6} = 1\frac{4}{6}$

    $\frac{6}{6}$ $\frac{4}{6}$

> I decompose $\frac{10}{6}$ into 2 parts: $\frac{6}{6}$ and $\frac{4}{6}$. $\frac{6}{6}$ is the same as 1, so I rewrite $\frac{10}{6}$ as the mixed number $1\frac{4}{6}$.

> I can think of the number sentence in unit form: 5 sixths + 5 sixths = 10 sixths.

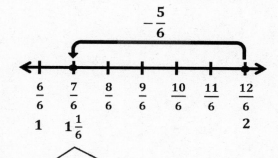

> I plot a point at $\frac{12}{6}$ because that is the whole. Then, I count backward to subtract $\frac{5}{6}$.

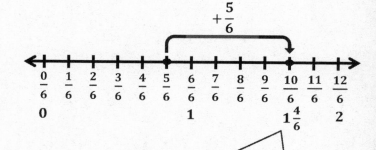

> I draw a number line and plot a point at $\frac{5}{6}$. I count up $\frac{5}{6}$. The model verifies the sum is $1\frac{4}{6}$.

Name _____ Date _____

1. Solve.

    a.   3 sixths − 2 sixths = _____

    b.   5 tenths − 3 tenths = _____

    c.   3 fourths − 2 fourths = _____

    d.   5 thirds − 2 thirds = _____

2. Solve.

    a. $\frac{3}{5} - \frac{2}{5}$

    b. $\frac{7}{9} - \frac{3}{9}$

    c. $\frac{7}{12} - \frac{3}{12}$

    d. $\frac{6}{6} - \frac{4}{6}$

    e. $\frac{5}{3} - \frac{2}{3}$

    f. $\frac{7}{4} - \frac{5}{4}$

3. Solve. Use a number bond to decompose the difference. Record your final answer as a mixed number. Problem (a) has been completed for you.

    a. $\frac{12}{6} - \frac{3}{6} = \frac{9}{6} = 1\frac{3}{6}$

    $\frac{6}{6} \qquad \frac{3}{6}$

    b. $\frac{17}{8} - \frac{6}{8}$

    c. $\frac{9}{5} - \frac{3}{5}$

    d. $\frac{11}{4} - \frac{6}{4}$

    e. $\frac{10}{7} - \frac{2}{7}$

    f. $\frac{21}{10} - \frac{9}{10}$

EUREKA MATH

Lesson 16:   Use visual models to add and subtract two fractions with the same units.

79

© 2018 Great Minds®. eureka-math.org

4. Solve. Write the sum in unit form.

   a. 4 fifths + 2 fifths = _____

   b. 5 eighths + 2 eighths = _____

5. Solve.

   a. $\dfrac{3}{11} + \dfrac{6}{11}$

   b. $\dfrac{3}{10} + \dfrac{6}{10}$

6. Solve. Use a number bond to decompose the sum. Record your final answer as a mixed number.

   a. $\dfrac{3}{4} + \dfrac{3}{4}$

   b. $\dfrac{8}{12} + \dfrac{6}{12}$

   c. $\dfrac{5}{8} + \dfrac{7}{8}$

   d. $\dfrac{8}{10} + \dfrac{5}{10}$

   e. $\dfrac{3}{5} + \dfrac{6}{5}$

   f. $\dfrac{4}{3} + \dfrac{2}{3}$

7. Solve. Use a number line to model your answer.

   a. $\dfrac{11}{9} - \dfrac{5}{9}$

   b. $\dfrac{13}{12} + \dfrac{4}{12}$

     Lesson 16:   Use visual models to add and subtract two fractions with the same units.

EUREKA
MATH®

1. Use the three fractions $\frac{8}{8}, \frac{3}{8}$, and $\frac{5}{8}$ to write two addition and two subtraction number sentences.

$$\frac{3}{8} + \frac{5}{8} = \frac{8}{8} \qquad\qquad \frac{8}{8} - \frac{5}{8} = \frac{3}{8}$$

$$\frac{5}{8} + \frac{3}{8} = \frac{8}{8} \qquad\qquad \frac{8}{8} - \frac{3}{8} = \frac{5}{8}$$

This is like the relationship between 3, 5, and 8:

$$3 + 5 = 8 \qquad 8 - 5 = 3$$
$$5 + 3 = 8 \qquad 8 - 3 = 5$$

except these fractions have units of eighths.

2. Solve by subtracting and counting up. Model with a number line.

$$1 - \frac{3}{8}$$

$$\frac{8}{8} - \frac{3}{8} = \frac{5}{8}$$

Or, I count up by thinking about how many eighths it takes to get from $\frac{3}{8}$ to $\frac{8}{8}$.

$$\frac{3}{8} + x = \frac{8}{8}$$
$$x = \frac{5}{8}$$

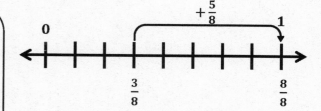

I rename 1 as $\frac{8}{8}$. Now, I have like units, eighths, and I can subtract.

A number line shows how to count up from $\frac{3}{8}$ to $\frac{8}{8}$. I can also start at 1 and show the subtraction of $\frac{3}{8}$ on the number line.

3. Find the difference in two ways. Use a number bond to decompose the whole.

$$1\frac{5}{8} - \frac{7}{8}$$

$$\frac{8}{8}, \frac{5}{8} \quad \frac{13}{8}$$

$$\frac{13}{8} - \frac{7}{8} = \boxed{\frac{6}{8}}$$

$$\frac{8}{8} \quad \frac{7}{8} - \frac{1}{8}$$

$$\frac{1}{8} + \frac{5}{8} = \boxed{\frac{6}{8}}$$

bond to rename $\frac{}{8}$ as $\frac{}{}$ and $\frac{}{}$.

I rename — as a fraction greater than . I have like units, so I can subtract — from —.

Or, I can subtract — from —, or , first and then add the remaining part of the number bond, —.

EUREKA MATH

Lesson 17:   Use visual models to add and subtract two fractions with the same units, including subtracting from one whole.

© 2018 Great Minds®. eureka-math.org

81

Name _____ Date _____

1. Use the following three fractions to write two subtraction and two addition number sentences.

a. $\frac{5}{6}$, $\frac{4}{6}$, $\frac{9}{6}$	b. $\frac{5}{9}$, $\frac{13}{9}$, $\frac{8}{9}$

2. Solve. Model each subtraction problem with a number line, and solve by both counting up and subtracting.

a. $1 - \frac{5}{8}$

b. $1 - \frac{2}{5}$

c. $1\frac{3}{6} - \frac{5}{6}$

d. $1 - \frac{1}{4}$

e. $1\frac{1}{3} - \frac{2}{3}$

f. $1\frac{1}{5} - \frac{2}{5}$

EUREKA
MATH

Lesson 17: Use visual models to add and subtract two fractions with the same units, including subtracting from one whole.

© 2018 Great Minds®. eureka-math.org

83

3. Find the difference in two ways. Use number bonds to decompose the total. Part (a) has been completed for you.

a. $1\frac{2}{5} - \frac{4}{5}$

$$\frac{5}{5} + \frac{2}{5} = \frac{7}{5}$$

$$\frac{7}{5} - \frac{4}{5} = \boxed{\frac{3}{5}}$$

$$\frac{5}{5} - \frac{4}{5} = \frac{1}{5}$$

$$\frac{1}{5} + \frac{2}{5} = \boxed{\frac{3}{5}}$$

b. $1\frac{3}{8} - \frac{7}{8}$

c. $1\frac{1}{4} - \frac{3}{4}$

d. $1\frac{2}{7} - \frac{5}{7}$

e. $1\frac{3}{10} - \frac{7}{10}$

EUREKA MATH

Show two ways to solve each problem. Express the answer as a mixed number when possible. Use a number bond when it helps you.

1. $\frac{2}{5} + \frac{3}{5} + \frac{1}{5}$

$$\frac{2}{5} + \frac{3}{5} = \frac{5}{5} = 1$$

$$1 + \frac{1}{5} = 1\frac{1}{5}$$

> I can add $\frac{2}{5}$ and $\frac{3}{5}$ to make 1. Then, I can just add $\frac{1}{5}$ more to get $1\frac{1}{5}$.

$$\frac{2}{5} + \frac{3}{5} + \frac{1}{5} = \frac{6}{5} = 1\frac{1}{5}$$

$$\frac{5}{5} \qquad \frac{1}{5}$$

> Since the units, or denominators, are the same for each addend, fifths, I can just add the number of units, or numerators.

> I can use a number bond to decompose $\frac{6}{5}$ into $\frac{5}{5}$ and $\frac{1}{5}$. Since $\frac{5}{5} = 1$, I can rewrite $\frac{6}{5}$ as $1\frac{1}{5}$.

2. $1 - \frac{3}{12} - \frac{4}{12}$

> I add $\frac{3}{12}$ and $\frac{4}{12}$ to get $\frac{7}{12}$. I need to subtract a total of $\frac{7}{12}$ from 1.

$$\frac{3}{12} + \frac{4}{12} = \frac{7}{12}$$

$$\frac{12}{12} - \frac{7}{12} = \frac{5}{12}$$

> I can rename 1 as $\frac{12}{12}$, and I can subtract $\frac{7}{12}$ from $\frac{12}{12}$.

$$\frac{12}{12} - \frac{3}{12} = \frac{9}{12}$$

$$\frac{9}{12} - \frac{4}{12} = \frac{5}{12}$$

> I rename 1 as $\frac{12}{12}$. Then, I subtract $\frac{3}{12}$, and finally I subtract $\frac{4}{12}$.

EUREKA MATH®

Name _____ Date _____

1. Show one way to solve each problem. Express sums and differences as a mixed number when possible. Use number bonds when it helps you. Part (a) is partially completed.

a. $\frac{1}{3} + \frac{2}{3} + \frac{1}{3}$ $= \frac{3}{3} + \frac{1}{3} = 1 + \frac{1}{3}$ $= \underline{\hspace{1cm}}$	b. $\frac{5}{8} + \frac{5}{8} + \frac{3}{8}$	c. $\frac{4}{6} + \frac{6}{6} + \frac{1}{6}$
d. $1\frac{2}{12} - \frac{2}{12} - \frac{1}{12}$	e. $\frac{5}{7} + \frac{1}{7} + \frac{4}{7}$	f. $\frac{4}{10} + \frac{7}{10} + \frac{9}{10}$
g. $1 - \frac{3}{10} - \frac{1}{10}$	h. $1\frac{3}{5} - \frac{4}{5} - \frac{1}{5}$	i. $\frac{10}{15} + \frac{7}{15} + \frac{12}{15} + \frac{1}{15}$

2. Bonnie used two different strategies to solve $\frac{5}{10} + \frac{4}{10} + \frac{3}{10}$.

Bonnie's First Strategy	**Bonnie's Second Strategy**

$$\frac{5}{10} + \frac{4}{10} + \frac{3}{10} = \frac{9}{10} + \frac{3}{10} = \frac{10}{10} + \frac{2}{10} = 1\frac{2}{10}$$

$$\frac{1}{10} \qquad \frac{2}{10}$$

$$\frac{5}{10} + \frac{4}{10} + \frac{3}{10} = \frac{12}{10} = 1 + \frac{2}{10} = 1\frac{2}{10}$$

$$\frac{10}{10} \qquad \frac{2}{10}$$

Which strategy do you like best? Why?

3. You gave one solution for each part of Problem 1. Now, for each problem indicated below, give a different solution method.

1(b) $\frac{5}{8} + \frac{5}{8} + \frac{3}{8}$

1(e) $\frac{5}{7} + \frac{1}{7} + \frac{4}{7}$

1(h) $1\frac{3}{5} - \frac{4}{5} - \frac{1}{5}$

Lesson 18:   Add and subtract more than two fractions.

EUREKA MATH

Use the RDW process to solve.

1. Noah drank $\frac{8}{10}$ liter of water on Monday and $\frac{6}{10}$ liter on Tuesday. How many liters of water did Noah drink in the 2 days?

w

$\frac{8}{10}$	$\frac{6}{10}$

I draw a tape diagram to model the problem. The parts in my tape diagram represent the water Noah drank on Monday and Tuesday. I use the variable w to represent the liters of water Noah drank on Monday and Tuesday.

$\frac{8}{10} + \frac{6}{10} = w$

I add the parts in my tape diagram to find the total amount of water that Noah drank.

$\frac{8}{10} + \frac{6}{10} = \frac{14}{10} = 1\frac{4}{10}$

$\frac{10}{10} \qquad \frac{4}{10}$

Since the addends have like units, I add the numerators to get $\frac{14}{10}$. I use a number bond to decompose $\frac{14}{10}$ into a whole number and a fraction. This helps me rename $\frac{14}{10}$ as a mixed number.

$w = 1\frac{4}{10}$

Noah drank $1\frac{4}{10}$ liters of water.

I write a statement to answer the question. I also think about the reasonableness of my answer. The water drunk on each day is less than 1 liter, so I would expect to get a total less than 2 liters. My answer of $1\frac{4}{10}$ liters is a reasonable total amount.

2. Muneeb had 2 chapters to read for homework. By 9:00 p.m., he had read $1\frac{2}{7}$ chapters. What fraction of chapters is left for Muneeb to read?

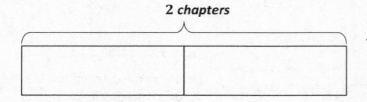

2 chapters

I can draw a tape diagram with 2 equal parts to represent the 2 chapters of the book.

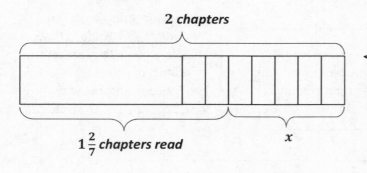

2 chapters

$1\frac{2}{7}$ **chapters read** x

To show $1\frac{2}{7}$ on my tape diagram, I partition one chapter into sevenths. I label the amount that Muneeb has read and the amount that is left, x.

$2 - 1\frac{2}{7} = x$

The unknown in my tape diagram is one of the parts, so I subtract the known part, $1\frac{2}{7}$, from the whole, 2.

$2 - 1\frac{2}{7} = \frac{5}{7}$

1            $\frac{7}{7}$

I use a number bond to show how to decompose one of the chapters into sevenths. My tape diagram shows that there is $\frac{5}{7}$ of a chapter left. My equation shows that, too!

$x = \frac{5}{7}$

Muneeb has $\frac{5}{7}$ chapter left to read.

Muneeb started with 2 chapters to read. He read 1 chapter and a little more, so he should have less than 1 chapter left. My answer of $\frac{5}{7}$ chapter is a reasonable amount left because it's less than 1 chapter.

Lesson 19:   Solve word problems involving addition and subtraction of fractions.   **EUREKA MATH**

Name _____ Date _____

Use the RDW process to solve.

1. Isla walked $\frac{3}{4}$ mile each way to and from school on Wednesday. How many miles did Isla walk that day?

2. Zach spent $\frac{2}{3}$ hour reading on Friday and $1\frac{1}{3}$ hours reading on Saturday. How much more time did he read on Saturday than on Friday?

3. Mrs. Cashmore bought a large melon. She cut a piece that weighed $1\frac{1}{8}$ pounds and gave it to her neighbor. The remaining piece of melon weighed $\frac{6}{8}$ pound. How much did the whole melon weigh?

Lesson 19: Solve word problems involving addition and subtraction of fractions.

91

4. Ally's little sister wanted to help her make some oatmeal cookies. First, she put $\frac{5}{8}$ cup of oatmeal in the bowl. Next, she added another $\frac{5}{8}$ cup of oatmeal. Finally, she added another $\frac{5}{8}$ cup of oatmeal. How much oatmeal did she put in the bowl?

5. Marcia baked 2 pans of brownies. Her family ate $1\frac{5}{6}$ pans. What fraction of a pan of brownies was left?

6. Joanie wrote a letter that was $1\frac{1}{4}$ pages long. Katie wrote a letter that was $\frac{3}{4}$ page shorter than Joanie's letter. How long was Katie's letter?

Lesson 19:   Solve word problems involving addition and subtraction of fractions.   EUREKA MATH

1. Use a tape diagram to represent each addend. Decompose one of the tape diagrams to make like units. Then, write the complete number sentence.

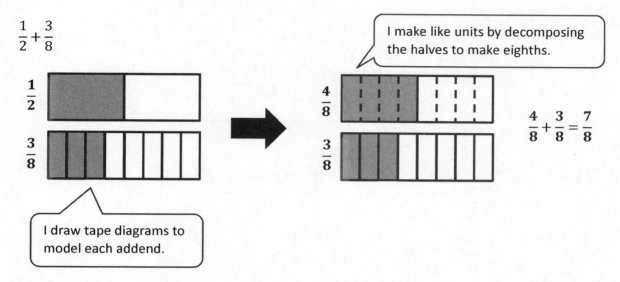

$\frac{1}{2} + \frac{3}{8}$

$\frac{1}{2}$

$\frac{3}{8}$

I make like units by decomposing the halves to make eighths.

$\frac{4}{8}$

$\frac{3}{8}$

$\frac{4}{8} + \frac{3}{8} = \frac{7}{8}$

I draw tape diagrams to model each addend.

2. Estimate to determine if the sum is between 0 and 1 or 1 and 2. Draw a number line to model the addition. Then, write a complete number sentence.

$\frac{7}{10} + \frac{1}{2}$

$\frac{7}{10}$ is a little bit more than $\frac{1}{2}$. When I add a fraction that is a little bigger than $\frac{1}{2}$ to $\frac{1}{2}$, I should get a total that is between 1 and 2.

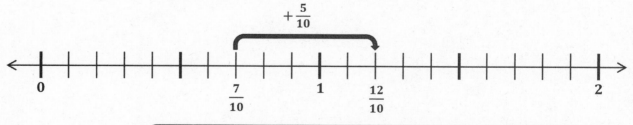

$+\frac{5}{10}$

0        $\frac{7}{10}$ 1 $\frac{12}{10}$ 2

$\frac{7}{10} + \frac{5}{10} = \frac{12}{10}$

To make like units in order to add, I decompose halves. The number line and the number sentence show the total, $\frac{12}{10}$, which is between 1 and 2.

Lesson 20:   Use visual models to add two fractions with related units using the denominators 2, 3, 4, 5, 6, 8, 10, and 12.   93

EUREKA MATH®

© 2018 Great Minds®. eureka-math.org

3. Solve the following addition problem without drawing a model. Show your work.

$\frac{2}{3} + \frac{1}{9}$

$\frac{2}{3} = \frac{2 \times 3}{3 \times 3} = \frac{6}{9}$

I can decompose thirds to make ninths by multiplying the numerator and denominator of $\frac{2}{3}$ by 3.

$\frac{6}{9} + \frac{1}{9} = \frac{7}{9}$

Now, I have like units, ninths, and I can add.

Lesson 20: Use visual models to add two fractions with related units using the denominators 2, 3, 4, 5, 6, 8, 10, and 12.

EUREKA MATH®

Name _____ Date _____

1. Use a tape diagram to represent each addend. Decompose one of the tape diagrams to make like units.
   Then, write the complete number sentence.

   a. $\frac{1}{3} + \frac{1}{6}$

   b. $\frac{1}{2} + \frac{1}{4}$

   c. $\frac{3}{4} + \frac{1}{8}$

   d. $\frac{1}{4} + \frac{5}{12}$

   e. $\frac{3}{8} + \frac{1}{2}$

   f. $\frac{3}{5} + \frac{3}{10}$

EUREKA
MATH®

Lesson 20: Use visual models to add two fractions with related units using the
denominators 2, 3, 4, 5, 6, 8, 10, and 12.

95

© 2018 Great Minds®. eureka-math.org

2. Estimate to determine if the sum is between 0 and 1 or 1 and 2. Draw a number line to model the addition. Then, write a complete number sentence. The first one has been completed for you.

a. $\frac{1}{3} + \frac{1}{6}$ $\frac{2}{6} + \frac{1}{6} = \frac{3}{6}$

b. $\frac{3}{5} + \frac{7}{10}$

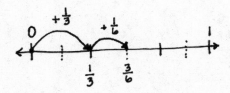

c. $\frac{5}{12} + \frac{1}{4}$

d. $\frac{3}{4} + \frac{5}{8}$

e. $\frac{7}{8} + \frac{3}{4}$

f. $\frac{1}{6} + \frac{5}{3}$

3. Solve the following addition problem without drawing a model. Show your work.

$$\frac{5}{6} + \frac{1}{3}$$

Lesson 20:   Use visual models to add two fractions with related units using the denominators 2, 3, 4, 5, 6, 8, 10, and 12.

© 2018 Great Minds®. eureka-math.org

EUREKA MATH

1. Use a tape diagram to represent each addend. Decompose one of the tape diagrams to make like units. Then, write the complete number sentence. Use a number bond to write the sum as a mixed number.

$\frac{5}{6} + \frac{2}{3}$

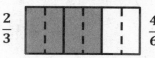

$\frac{5}{6}$

$\frac{2}{3}$ $\frac{4}{6}$

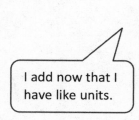

$\frac{5}{6} + \frac{4}{6} = \frac{9}{6} = 1\frac{3}{6}$

$\frac{6}{6} \qquad \frac{3}{6}$

I add now that I have like units.

I can make like units by decomposing the thirds as sixths. I decompose the thirds because they are the larger unit (thirds > sixths).

2. Draw a number line to model the addition. Then, write a complete number sentence. Use a number bond to write the sum as a mixed number.

$\frac{1}{2} + \frac{7}{8}$ $\frac{1}{2} = \frac{1 \times 4}{2 \times 4} = \frac{4}{8}$

I rename halves as eighths to make like units to add.

$\frac{4}{8} + \frac{7}{8} = \frac{11}{8} = 1\frac{3}{8}$

$\frac{8}{8} \qquad \frac{3}{8}$

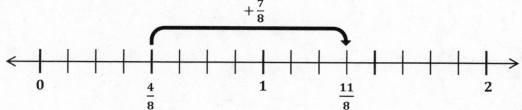

$+\frac{7}{8}$

0      $\frac{4}{8}$ 1 $\frac{11}{8}$ 2

3. Solve. Write the sum as a mixed number. Draw a model if needed.

$\frac{5}{6} + \frac{2}{3}$

$\frac{5}{6} + \frac{2}{3} = \frac{5}{6} + \frac{4}{6} = \frac{9}{6} = 1\frac{3}{6}$

$\frac{6}{6} \qquad \frac{3}{6}$

I double the units (denominator) to make sixths, which means I also need to double the number of units (numerator). $\frac{2}{3}$ is equal to $\frac{4}{6}$.

EUREKA MATH®

Lesson 21:   Use visual models to add two fractions with related units using the denominators 2, 3, 4, 5, 6, 8, 10, and 12.

Name _____   Date _____

1. Draw a tape diagram to represent each addend. Decompose one of the tape diagrams to make like units. Then, write a complete number sentence. Use a number bond to write each sum as a mixed number.

a. $\frac{7}{8} + \frac{1}{4}$ b. $\frac{4}{8} + \frac{2}{4}$

c. $\frac{4}{6} + \frac{1}{2}$ d. $\frac{3}{5} + \frac{8}{10}$

2. Draw a number line to model the addition. Then, write a complete number sentence. Use a number bond to write each sum as a mixed number.

a. $\frac{1}{2} + \frac{5}{8}$ b. $\frac{3}{4} + \frac{3}{8}$

EUREKA MATH®   **Lesson 21:**   Use visual models to add two fractions with related units using the denominators 2, 3, 4, 5, 6, 8, 10, and 12.   99

© 2018 Great Minds®. eureka-math.org

c. $\frac{4}{10} + \frac{4}{5}$

d. $\frac{1}{3} + \frac{5}{6}$

3. Solve. Write the sum as a mixed number. Draw a model if needed.

a. $\frac{1}{2} + \frac{6}{8}$

b. $\frac{7}{8} + \frac{3}{4}$

c. $\frac{5}{6} + \frac{1}{3}$

d. $\frac{9}{10} + \frac{2}{5}$

e. $\frac{4}{12} + \frac{3}{4}$

f. $\frac{1}{2} + \frac{5}{6}$

g. $\frac{3}{12} + \frac{5}{6}$

h. $\frac{7}{10} + \frac{4}{5}$

Lesson 21: Use visual models to add two fractions with related units using the denominators 2, 3, 4, 5, 6, 8, 10, and 12.

EUREKA MATH

1. Draw a tape diagram to match the number sentence. Then, complete the number sentence.

$3 - \dfrac{2}{4} = \underline{2\dfrac{2}{4}}$

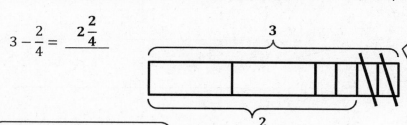

3

$2\dfrac{2}{4}$

I draw a tape diagram with 3 equal units, with 1 unit decomposed into fourths. To show the subtraction, I cross off $\dfrac{2}{4}$.

The tape diagram shows the difference is $2\dfrac{2}{4}$.

2. Use $\dfrac{5}{6}$, 3, and $2\dfrac{1}{6}$ to write two subtraction and two addition number sentences.

$\dfrac{5}{6} + 2\dfrac{1}{6} = 3$

$2\dfrac{1}{6} + \dfrac{5}{6} = 3$

$3 - \dfrac{5}{6} = 2\dfrac{1}{6}$

$3 - 2\dfrac{1}{6} = \dfrac{5}{6}$

I can also represent the relationship between these 3 numbers with a number bond.

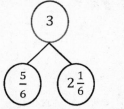

3. Solve using a number bond. Draw a number line to represent the number sentence.

$4 - \dfrac{2}{3} = \underline{3\dfrac{1}{3}}$

 $4 - \dfrac{2}{3} = 3\dfrac{1}{3}$

3   $\dfrac{3}{3}$

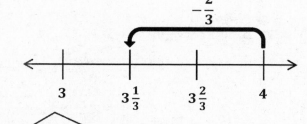

$-\dfrac{2}{3}$

3   $3\dfrac{1}{3}$ $3\dfrac{2}{3}$ 4

I use a number bond to decompose 4 into 3 and $\dfrac{3}{3}$. Then, I subtract $\dfrac{2}{3}$ from $\dfrac{3}{3}$.

I draw a number line with the endpoints 3 and 4 because I am starting at 4 and subtracting a number less than 1.

EUREKA MATH

Lesson 22: Add a fraction less than 1 to, or subtract a fraction less than 1 from, a whole number using decomposition and visual models.

101

© 2018 Great Minds®. eureka-math.org

4. Complete the subtraction sentence using a number bond.

$$6 - \frac{6}{8} = \underline{5\frac{2}{8}}$$

Number bond: 6 decomposes into 5 and $\frac{8}{8}$.

$$\frac{8}{8} - \frac{6}{8} = \frac{2}{8}$$

$$5 + \frac{2}{8} = 5\frac{2}{8}$$

> I subtract $\frac{6}{8}$ from $\frac{8}{8}$ to get $\frac{2}{8}$. I add $\frac{2}{8}$ back to 5.

Lesson 22:   Add a fraction less than 1 to, or subtract a fraction less than 1 from, a whole number using decomposition and visual models.

© 2018 Great Minds®. eureka-math.org

EUREKA MATH®

Name _____   Date _____

1. Draw a tape diagram to match each number sentence. Then, complete the number sentence.

   a. $2 + \dfrac{1}{4} = $ _____

   b. $3 + \dfrac{2}{3} = $ _____

   c. $2 - \dfrac{1}{5} = $_____

   d. $3 - \dfrac{3}{4} = $ _____

2. Use the following three numbers to write two subtraction and two addition number sentences.

   a. $4,\ 4\dfrac{5}{8},\ \dfrac{5}{8}$

   b. $\dfrac{2}{7},\ 5\dfrac{5}{7},\ 6$

3. Solve using a number bond. Draw a number line to represent each number sentence. The first one has been done for you.

   a. $4 - \dfrac{1}{3} = \quad 3\dfrac{2}{3}$

   b. $8 - \dfrac{5}{6} = $ _____

   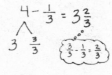

   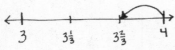

EUREKA MATH   **Lesson 22:**   Add a fraction less than 1 to, or subtract a fraction less than 1 from, a whole number using decomposition and visual models.   **103**

© 2018 Great Minds®. eureka-math.org

c. $7 - \frac{4}{5} =$ _____

d. $3 - \frac{3}{10} =$ _____

4. Complete the subtraction sentences using number bonds.

a. $6 - \frac{1}{4} =$ _____

b. $7 - \frac{2}{10} =$ _____

c. $5 - \frac{5}{6} =$ _____

d. $6 - \frac{6}{8} =$ _____

e. $3 - \frac{7}{8} =$ _____

f. $26 - \frac{7}{10} =$ _____

Lesson 22: Add a fraction less than 1 to, or subtract a fraction less than 1 from, a whole number using decomposition and visual models.

© 2018 Great Minds®. eureka-math.org

EUREKA MATH

1. Count by 1 fifths. Start at 0 fifths. End at 10 fifths. Circle any fractions that are equivalent to a whole number. Record the whole number below the fraction.

$$\left(\frac{0}{5}\right), \ \frac{1}{5}, \ \frac{2}{5}, \ \frac{3}{5}, \ \frac{4}{5}, \ \left(\frac{5}{5}\right), \ \frac{6}{5}, \ \frac{7}{5}, \ \frac{8}{5}, \ \frac{9}{5}, \ \left(\frac{10}{5}\right)$$

  0   1   2

> I know that 5 fifths equals 1, so 10 fifths equals 2.

2. Use parentheses to show how to make ones in the following number sentence.

$$\left(\frac{1}{4}+\frac{1}{4}+\frac{1}{4}+\frac{1}{4}\right)+\left(\frac{1}{4}+\frac{1}{4}+\frac{1}{4}+\frac{1}{4}\right)=2$$

> I draw parentheses around groups of 4 fourths because the denominator (fourths) tells me how many unit fractions composed make 1.

3. Multiply. Draw a number line to support your answer.

$$4 \times \frac{1}{2}$$

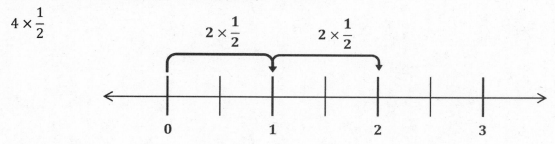

$$4 \times \frac{1}{2} = 2 \times \frac{2}{2} = 2$$

> I see on my number line that 4 copies of $\frac{1}{2}$ is the same as 2 copies of $\frac{2}{2}$. Since $\frac{2}{2}$ is the same as 1, I think of 2 copies of $\frac{2}{2}$ as the multiplication sentence, $2 \times 1 = 2$. So, $4 \times \frac{1}{2} = 2$.

4. Multiply. Write the product as a mixed number. Draw a number line to support your answer.

$11 \times \dfrac{1}{4}$

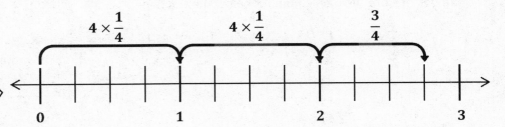

I draw a number line and partition each whole into fourths since the fractional unit that I'm multiplying by is fourths.

$$11 \times \frac{1}{4} = \left(2 \times \frac{4}{4}\right) + \frac{3}{4} = 2 + \frac{3}{4} = 2\frac{3}{4}$$

I can see on my number line that 11 copies of $\frac{1}{4}$ equals 2 copies of $\frac{4}{4}$ plus $\frac{3}{4}$.

Lesson 23: Add and multiply unit fractions to build fractions greater than 1 using visual models.

EUREKA MATH

Name _____ Date _____

1. Circle any fractions that are equivalent to a whole number. Record the whole number below the fraction.

   a. Count by 1 fourths. Start at 0 fourths. Stop at 6 fourths.

   $\dfrac{0}{4},$ $\dfrac{1}{4},$

   0

   b. Count by 1 sixths. Start at 0 sixths. Stop at 14 sixths.

2. Use parentheses to show how to make ones in the following number sentence.

$$\frac{1}{3}+\frac{1}{3}+\frac{1}{3}+\frac{1}{3}+\frac{1}{3}+\frac{1}{3}+\frac{1}{3}+\frac{1}{3}+\frac{1}{3}+\frac{1}{3}+\frac{1}{3}+\frac{1}{3}=4$$

3. Multiply, as shown below. Draw a number line to support your answer.

   a. $6 \times \dfrac{1}{3}$

   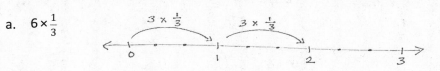

   $$6 \times \frac{1}{3} = 2 \times \frac{3}{3} = 2$$

   b. $10 \times \dfrac{1}{2}$

   c. $8 \times \dfrac{1}{4}$

EUREKA MATH

Lesson 23:   Add and multiply unit fractions to build fractions greater than 1 using visual models.

107

© 2018 Great Minds®. eureka-math.org

4. Multiply, as shown below. Write the product as a mixed number. Draw a number line to support your answer.

a. 7 copies of 1 third

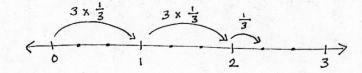

$$7 \times \frac{1}{3} = \left(2 \times \frac{3}{3}\right) + \frac{1}{3} = 2 + \frac{1}{3} = 2\frac{1}{3}$$

b. 7 copies of 1 fourth

c. 11 groups of 1 fifth

d. $7 \times \frac{1}{2}$

e. $9 \times \frac{1}{5}$

Lesson 23:   Add and multiply unit fractions to build fractions greater than 1 using visual models.

© 2018 Great Minds®. eureka-math.org

EUREKA MATH®

1. Rename $\frac{10}{3}$ as a mixed number by decomposing it into two parts. Model the decomposition with a number line and a number bond.

$$\frac{10}{3} = \frac{9}{3} + \frac{1}{3} = 3 + \frac{1}{3} = 3\frac{1}{3}$$

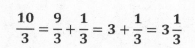

$\frac{9}{3}$ $\frac{1}{3}$

> I choose the 2 parts $\frac{9}{3}$ and $\frac{1}{3}$ for the number bond because $\frac{9}{3}$ is 3 groups of $\frac{3}{3}$, or 3. Then, I add the other part of my number bond, $\frac{1}{3}$, to get the mixed number $3\frac{1}{3}$.

> The number line shows that decomposing $\frac{10}{3}$ as $\frac{9}{3}$ and $\frac{1}{3}$ is the same as $3\frac{1}{3}$.

2. Rename $\frac{8}{3}$ as a mixed number using multiplication. Draw a number line to support your answer.

$$\frac{8}{3} = \frac{3 \times 2}{3} + \frac{2}{3} = 2 + \frac{2}{3} = 2\frac{2}{3}$$

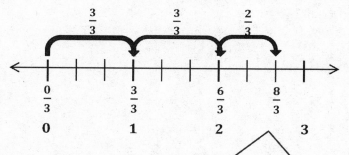

> I use multiplication to show that $\frac{6}{3}$ is 2 copies of $\frac{3}{3}$, which is the same as 2.

> The number line supports $\frac{8}{3}$ renamed as $2\frac{2}{3}$. They are equal.

3. Convert $\frac{22}{7}$ to a mixed number.

$$\frac{22}{7} = \left(3 \times \frac{7}{7}\right) + \frac{1}{7} = 3 + \frac{1}{7} = 3\frac{1}{7}$$

> I can make 3 groups of $\frac{7}{7}$, which equals $\frac{21}{7}$.
> I can add 1 more seventh to equal $\frac{22}{7}$.

EUREKA MATH®

Name _____ Date _____

1. Rename each fraction as a mixed number by decomposing it into two parts as shown below. Model the decomposition with a number line and a number bond.

   a. $\frac{11}{3}$

   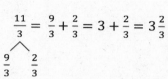

   $$\frac{11}{3} = \frac{9}{3} + \frac{2}{3} = 3 + \frac{2}{3} = 3\frac{2}{3}$$

   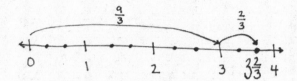

   b. $\frac{13}{4}$

   c. $\frac{16}{5}$

   d. $\frac{15}{2}$

   e. $\frac{17}{3}$

EUREKA MATH

Lesson 24:   Decompose and compose fractions greater than 1 to express them in various forms.

111

© 2018 Great Minds®. eureka-math.org

2. Convert each fraction to a mixed number. Show your work as in the example. Model with a number line.

    a. $\frac{11}{3}$

$$\frac{11}{3} = \frac{3 \times 3}{3} + \frac{2}{3} = 3 + \frac{2}{3} = 3\frac{2}{3}$$

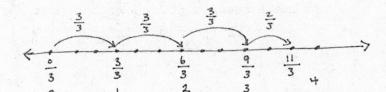

    b. $\frac{13}{2}$

    c. $\frac{18}{4}$

3. Convert each fraction to a mixed number.

a. $\frac{14}{3}$ =	b. $\frac{17}{4}$ =	c. $\frac{27}{5}$ =
d. $\frac{28}{6}$ =	e. $\frac{23}{7}$ =	f. $\frac{37}{8}$ =
g. $\frac{51}{9}$ =	h. $\frac{74}{10}$ =	i. $\frac{45}{12}$ =

Lesson 24:   Decompose and compose fractions greater than 1 to express them in various forms.

© 2018 Great Minds®. eureka-math.org

EUREKA MATH

1. Convert the mixed number $2\frac{2}{4}$ to a fraction greater than 1. Draw a number line to model your work.

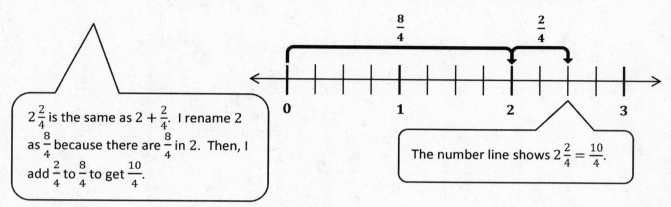

$2\frac{2}{4}$ is the same as $2 + \frac{2}{4}$. I rename 2 as $\frac{8}{4}$ because there are $\frac{8}{4}$ in 2. Then, I add $\frac{2}{4}$ to $\frac{8}{4}$ to get $\frac{10}{4}$.

The number line shows $2\frac{2}{4} = \frac{10}{4}$.

2. Use multiplication to convert the mixed number $5\frac{1}{4}$ to a fraction greater than 1.

$$5\frac{1}{4} = 5 + \frac{1}{4} = \left(5 \times \frac{4}{4}\right) + \frac{1}{4} = \frac{20}{4} + \frac{1}{4} = \frac{21}{4}$$

I rewrite 5 as the multiplication expression, $5 \times \frac{4}{4}$. Then, I can multiply $5 \times \frac{4}{4}$ to get $\frac{20}{4}$. So, there are $\frac{20}{4}$ in 5. Then, I add the $\frac{1}{4}$ from the $5\frac{1}{4}$ to get $\frac{21}{4}$.

3. Convert the mixed number $6\frac{1}{3}$ to a fraction greater than 1.

$$6\frac{1}{3} = \frac{18}{3} + \frac{1}{3} = \frac{19}{3}$$

I use mental math. There are 6 ones and 1 third in the number $6\frac{1}{3}$. I know that there are 18 thirds in 6 ones. 18 thirds plus 1 more third is 19 thirds.

Name _____ Date _____

1. Convert each mixed number to a fraction greater than 1. Draw a number line to model your work.

a. $3\frac{1}{4}$

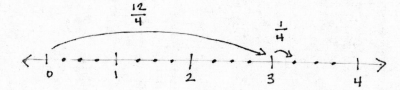

$$3\frac{1}{4} = 3 + \frac{1}{4} = \frac{12}{4} + \frac{1}{4} = \frac{13}{4}$$

b. $4\frac{2}{5}$

c. $5\frac{3}{8}$

d. $3\frac{7}{10}$

e. $6\frac{2}{9}$

EUREKA MATH

Lesson 25: Decompose and compose fractions greater than 1 to express them in various forms.

115

© 2018 Great Minds®. eureka-math.org

2. Convert each mixed number to a fraction greater than 1. Show your work as in the example.

(Note: $3 \times \frac{4}{4} = \frac{3 \times 4}{4}$.)

a. $3\frac{3}{4}$

$$3\frac{3}{4} = 3 + \frac{3}{4} = \left(3 \times \frac{4}{4}\right) + \frac{3}{4} = \frac{12}{4} + \frac{3}{4} = \frac{15}{4}$$

b. $5\frac{2}{3}$

c. $4\frac{1}{5}$

d. $3\frac{7}{8}$

3. Convert each mixed number to a fraction greater than 1.

a. $2\frac{1}{3}$	b. $2\frac{3}{4}$	c. $3\frac{2}{5}$
d. $3\frac{1}{6}$	e. $4\frac{5}{12}$	f. $4\frac{2}{5}$
g. $4\frac{1}{10}$	h. $5\frac{1}{5}$	i. $5\frac{5}{6}$
j. $6\frac{1}{4}$	k. $7\frac{1}{2}$	l. $7\frac{11}{12}$

Lesson 25: Decompose and compose fractions greater than 1 to express them in various forms.

EUREKA MATH

1.

   a. Plot the following points on the number line without measuring.

   i. $6\frac{7}{8}$ ii. $\frac{36}{5} = 7\frac{1}{5}$ iii. $\frac{19}{3} = 6\frac{1}{3}$

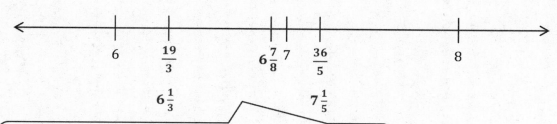

   To plot the numbers on the number line, I rewrite $\frac{36}{5}$ and $\frac{19}{3}$ as mixed numbers.

   I estimate to plot each number on the number line. I know that $6\frac{7}{8}$ is $\frac{1}{8}$ less than 7. I use this strategy to plot $6\frac{1}{3}$ and $7\frac{1}{5}$.

   b. Use the number line in Part 1(a) to compare the numbers by writing >, <, or =.

   i. $\frac{19}{3}$ ___<___ $6\frac{7}{8}$ ii. $\frac{36}{5}$ ___>___ $\frac{19}{3}$

   I remember from Lessons 12 and 13 how I used the benchmarks of 0, $\frac{1}{2}$, and 1 to compare. $\frac{19}{3}$ is less than $6\frac{1}{2}$, and $6\frac{7}{8}$ is greater than $6\frac{1}{2}$. $\frac{36}{5}$ is greater than 7 and $\frac{19}{3}$ is less than 7.

2. Compare the fractions given below by writing $>$, $<$, or $=$. Give a brief explanation for each answer, referring to benchmark fractions.

a. $4\frac{4}{8}$ ___$>$___ $4\frac{2}{5}$

$4\frac{4}{8}$ is the same as $4\frac{1}{2}$. $4\frac{2}{5}$ is less than $4\frac{1}{2}$, so $4\frac{4}{8}$ is greater than $4\frac{2}{5}$.

b. $\frac{43}{9}$ ___$<$___ $\frac{35}{7}$

$\frac{35}{7}$ is the same as 5. $\frac{43}{9}$ needs 2 more ninths to equal 5. That means that $\frac{35}{7}$ is greater than $\frac{43}{9}$.

Lesson 26: Compare fractions greater than 1 by reasoning using benchmark fractions.

© 2018 Great Minds®. eureka-math.org

EUREKA MATH

Name _____   Date _____

1. a.   Plot the following points on the number line without measuring.

   i. $2\frac{1}{6}$ ii. $3\frac{3}{4}$ iii. $\frac{33}{9}$

   <---|----------------------------|----------------------------|--->
       2   3   4

   b.   Use the number line in Problem 1(a) to compare the fractions by writing >, <, or =.

   i. $\frac{33}{9}$ _____ $2\frac{1}{6}$ ii. $\frac{33}{9}$ _____ $3\frac{3}{4}$

2. a.   Plot the following points on the number line without measuring.

   i. $\frac{65}{8}$ ii. $8\frac{5}{6}$ iii. $\frac{29}{4}$

   <---|----------------------------|----------------------------|--->
       7   8   9

   b.   Compare the following by writing >, <, or =.

   i. $8\frac{5}{6}$ _____ $\frac{65}{8}$ ii. $\frac{29}{4}$ _____ $\frac{65}{8}$

   c.   Explain how you plotted the points in Problem 2(a).

EUREKA
MATH

Lesson 26:   Compare fractions greater than 1 by reasoning using benchmark
fractions.

119

© 2018 Great Minds®. eureka-math.org

3. Compare the fractions given below by writing $>$, $<$, or $=$. Give a brief explanation for each answer, referring to benchmark fractions.

a. $5\frac{1}{3}$ _____ $5\frac{3}{4}$

b. $\frac{12}{4}$ _____ $\frac{25}{8}$

c. $\frac{18}{6}$ _____ $\frac{17}{4}$

d. $5\frac{3}{5}$ _____ $5\frac{5}{10}$

e. $6\frac{3}{4}$ _____ $6\frac{3}{5}$

f. $\frac{33}{6}$ _____ $\frac{34}{7}$

g. $\frac{23}{10}$ _____ $\frac{20}{8}$

h. $\frac{27}{12}$ _____ $\frac{15}{6}$

i. $2\frac{49}{50}$ _____ $2\frac{99}{100}$

j. $6\frac{5}{9}$ _____ $6\frac{49}{100}$

Lesson 26:   Compare fractions greater than 1 by reasoning using benchmark fractions.

EUREKA MATH®

1. Draw a tape diagram to model the comparison. Use >, <, or = to compare.

$5\frac{7}{8}$ ___>___ $\frac{23}{4}$

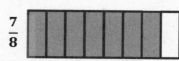

$\frac{7}{8}$

$\frac{23}{4} = 5\frac{3}{4}$

$\frac{3}{4} = \frac{6}{8}$

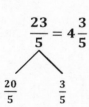

$\frac{20}{4}$ $\frac{3}{4}$

I can rename $\frac{23}{4}$ as a mixed number, $5\frac{3}{4}$.

Since both numbers have 5 ones, I draw tape diagrams to represent the fractional parts of each number. I decompose fourths to eighths. My tape diagrams show that $\frac{3}{4} = \frac{6}{8}$ and $\frac{7}{8} > \frac{6}{8}$.

2. Use an area model to make like units. Then, use >, <, or = to compare.

$4\frac{2}{3}$ ___>___ $\frac{23}{5}$

$\frac{23}{5} = 4\frac{3}{5}$

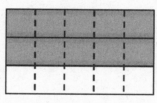

$\frac{20}{5}$ $\frac{3}{5}$

$\frac{2}{3} = \frac{10}{15}$

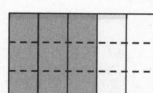

$\frac{3}{5} = \frac{9}{15}$

I draw area models to represent the fractional parts of each number. I make like units by drawing fifths vertically on the thirds and thirds horizontally on the fifths.

EUREKA MATH

Lesson 27: Compare fractions greater than 1 by creating common numerators or denominators.

121

© 2018 Great Minds®. eureka-math.org

3. Compare each pair of fractions using >, <, or = using any strategy.

a. $\dfrac{14}{6}$ __>__ $\dfrac{14}{9}$

Both fractions have the same numerator. Since sixths are bigger than ninths, $\dfrac{14}{6} > \dfrac{14}{9}$.

b. $\dfrac{19}{4}$ __<__ $\dfrac{25}{5}$

$\dfrac{25}{5} = 5$, and $\dfrac{19}{4} < 5$ because is takes 20 fourths to equal 5.

c. $6\dfrac{2}{6}$ __>__ $6\dfrac{4}{9}$

$\dfrac{2 \times 3}{6 \times 3} = \dfrac{6}{18}$

$\dfrac{4 \times 2}{9 \times 2} = \dfrac{8}{18}$

$\dfrac{6}{18} < \dfrac{8}{18}$

I make like units, eighteenths, and compare.

Lesson 27:   Compare fractions greater than 1 by creating common numerators or denominators.

EUREKA MATH®

Name _____   Date _____

1. Draw a tape diagram to model each comparison. Use >, <, or = to compare.

   a. $2\frac{3}{4}$ _____ $2\frac{7}{8}$ b. $10\frac{2}{6}$ _____ $10\frac{1}{3}$

   c. $5\frac{3}{8}$ _____ $5\frac{1}{4}$ d. $2\frac{5}{9}$ _____ $\frac{21}{3}$

2. Use an area model to make like units. Then, use >, <, or = to compare.

   a. $2\frac{4}{5}$ _____ $\frac{11}{4}$ b. $2\frac{3}{5}$ _____ $2\frac{2}{3}$

3. Compare each pair of fractions using >, <, or = using any strategy.

a. $6\frac{1}{2}$ _____ $6\frac{3}{8}$

b. $7\frac{5}{6}$ _____ $7\frac{11}{12}$

c. $3\frac{6}{10}$ _____ $3\frac{2}{5}$

d. $2\frac{2}{5}$ _____ $2\frac{8}{15}$

e. $\frac{10}{3}$ _____ $\frac{10}{4}$

f. $\frac{12}{4}$ _____ $\frac{10}{3}$

g. $\frac{38}{9}$ _____ $4\frac{2}{12}$

h. $\frac{23}{4}$ _____ $5\frac{2}{3}$

i. $\frac{30}{8}$ _____ $3\frac{7}{12}$

j. $10\frac{3}{4}$ _____ $10\frac{4}{6}$

Lesson 27:   Compare fractions greater than 1 by creating common numerators or denominators.

EUREKA MATH®

1. A group of students recorded the amount of time they spent doing homework in a week. The times are shown in the table. Make a line plot to display the data.

Student	Time Spent Doing Homework (in hours)	
Rebecca	$6\frac{1}{4}$	✓
Noah	6	✓
Wilson	$5\frac{3}{4}$	✓
Jenna	$6\frac{1}{4}$	✓
Sam	$6\frac{1}{2}$	✓
Angie	6	✓
Matthew	$6\frac{1}{4}$	✓
Jessica	$6\frac{3}{4}$	✓

I can make a line plot with an interval of fourths because that's the smallest unit in the table. My endpoints are $5\frac{3}{4}$ and $6\frac{3}{4}$ because those are the shortest and longest times spent doing homework. I can draw an X above the correct time on the number line to represent the time each student spent doing homework.

Time Spent Doing Homework in One Week

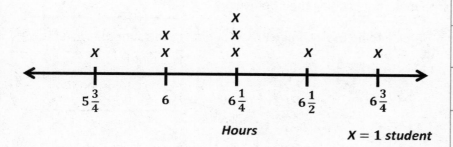

Hours   X = 1 student

2. Solve each problem.

a. Who spent 1 hour longer doing homework than Wilson?

$$5\frac{3}{4} + 1 = 6\frac{3}{4}$$

I can add 1 hour to Wilson's time and look at the table to find the answer.

Jessica spent 1 hour longer doing homework than Wilson.

b. How many quarter hours did Jenna spend doing homework?

$$6\frac{1}{4} = \frac{24}{4} + \frac{1}{4} = \frac{25}{4}$$

Jenna spent 25 quarter hours doing her homework.

c. What is the difference, in hours, between the most frequent amount of time spent doing homework and the second most frequent amount of time spent doing homework?

$$6\frac{1}{4} - 6 = \frac{1}{4}$$

The difference is 1 fourth hour.

> The X's on the line plot help me see the most frequent time, $6\frac{1}{4}$ hours, and the second most frequent time, 6 hours.

d. Compare the times of Matthew and Sam using >, <, or =.

$$6\frac{1}{4} < 6\frac{1}{2}$$

Matthew spent less time doing his homework than Sam.

e. How many students spent less than $6\frac{1}{2}$ hours doing their homework?

Six students spent less than $6\frac{1}{2}$ hours doing their homework.

> I can count the X's on the line plot for $5\frac{3}{4}$ hours, 6 hours, and $6\frac{1}{4}$ hours.

f. How many students recorded the amount of time they spent doing their homework?

Eight students recorded the amount of time they spent doing their homework.

> I can count the X's on the line plot, or I can count the students in the table.

g. Scott spent $\frac{30}{4}$ hours in one week doing his homework. Use >, <, or = to compare Scott's time to the time of the student who spent the most hours doing homework. Who spent more time doing homework?

$$\frac{30}{4} = \frac{28}{4} + \frac{2}{4} = 7 + \frac{2}{4} = 7\frac{2}{4}$$

$$7\frac{2}{4} > 6\frac{3}{4}$$

> I can rename Scott's time as a mixed number, and then I can compare (or I can rename Jessica's time as a fraction greater than 1). There are 7 ones in Scott's time and only 6 ones in Jessica's time.

Scott spent more time than Jessica doing homework.

     Lesson 28:   Solve word problems with line plots.

EUREKA MATH

Name _____ Date _____

1. A group of students measured the lengths of their shoes. The measurements are shown in the table. Make a line plot to display the data.

Students	Length of shoe (in inches)
Collin	$8\frac{1}{2}$
Dickon	$7\frac{3}{4}$
Ben	$7\frac{1}{2}$
Martha	$7\frac{3}{4}$
Lilias	8
Susan	$8\frac{1}{2}$
Frances	$7\frac{3}{4}$
Mary	$8\frac{3}{4}$

2. Solve each problem.

   a. Who has a shoe length 1 inch longer than Dickon's?

   b. Who has a shoe length 1 inch shorter than Susan's?

c.   How many quarter inches long is Martha's shoe length?

d.   What is the difference, in inches, between Lilias's and Martha's shoe lengths?

e.   Compare the shoe length of Ben and Frances using >, <, or =.

f.   How many students had shoes that measured less than 8 inches?

g.   How many students measured the length of their shoes?

h.   Mr. Jones's shoe length was $\frac{25}{2}$ inches. Use >, <, or = to compare the length of Mr. Jones's shoe to the length of the longest student shoe length. Who had the longer shoe?

3.   Using the information in the table and on the line plot, write a question you could solve by using the line plot. Solve.

EUREKA
MATH

1. Estimate each sum or difference to the nearest half or whole number by rounding. Explain your estimate using words or a number line.

   a. $4\frac{1}{9} + 2\frac{4}{5} \approx$ ___7___

   $4\frac{1}{9}$ is close to 4, and $2\frac{4}{5}$ is close to 3. $4 + 3 = 7$

   > $4\frac{1}{9}$ is 1 ninth more than 4. $2\frac{4}{5}$ is 1 fifth less than 3.

   b. $7\frac{5}{6} - 2\frac{1}{4} \approx$ ___6___

   > I draw a number line and plot the mixed numbers. It's easy to see on my number line that $7\frac{5}{6}$ is close to 8 and $2\frac{1}{4}$ is close to 2.

   estimated difference $8 - 2 = 6$

   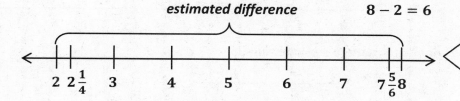

   > My number line makes it easy to see that the estimated difference is larger than the actual difference because I rounded one number up and the other number down.

   c. $5\frac{4}{10} + 3\frac{1}{8} \approx$ ___$8\frac{1}{2}$___

   $5\frac{4}{10}$ is close to $5\frac{1}{2}$, and $3\frac{1}{8}$ is close to 3. $5\frac{1}{2} + 3 = 8\frac{1}{2}$

   d. $\frac{15}{7} + \frac{20}{3} \approx$ ___9___ $\frac{15}{7} = 2\frac{1}{7}$ $\frac{20}{3} = 6\frac{2}{3}$

   $2 + 7 = 9$ $2\frac{1}{7} \approx 2$ $6\frac{2}{3} \approx 7$

   I renamed each fraction greater than 1 as a mixed number. Then, I rounded to the nearest whole number and added the rounded numbers.

2. Ben's estimate for $8\frac{6}{10} - 3\frac{1}{4}$ was 6. Michelle's estimate was $5\frac{1}{2}$. Whose estimate do you think is closer to the actual difference? Explain.

I think Michelle's estimate is closer to the actual difference. Ben rounded both numbers to the nearest whole number and then subtracted: $9 - 3 = 6$. Michelle rounded $8\frac{6}{10}$ to the nearest half, $8\frac{1}{2}$, and she rounded $3\frac{1}{4}$ to the nearest whole number. Then, she subtracted: $8\frac{1}{2} - 3 = 5\frac{1}{2}$. Since $8\frac{6}{10}$ is closer to $8\frac{1}{2}$ than 9, rounding it to the nearest half will give a closer estimate than rounding both numbers to the nearest whole number.

I can also draw number lines to show the actual difference, Ben's estimated difference, and Michelle's estimated difference. Because Ben rounded the total up and the part down, his estimated difference will be greater than the actual difference.

3. Use benchmark numbers or mental math to estimate the sum.

$14\frac{3}{8} + 7\frac{7}{12} \approx \textbf{22}$

$\mathbf{14\frac{1}{2} + 7\frac{1}{2} = 21 + 1 = 22}$

$\frac{3}{8}$ is 1 eighth less than $\frac{1}{2}$, and $\frac{7}{12}$ is 1 twelfth greater than $\frac{1}{2}$. I add the ones, and then I add the halves to get 22.

Lesson 29:   Estimate sums and differences using benchmark numbers.

EUREKA MATH

Name _____   Date _____

1. Estimate each sum or difference to the nearest half or whole number by rounding. Explain your estimate using words or a number line.

    a. $3\frac{1}{10} + 1\frac{3}{4} \approx$ _____

    b. $2\frac{9}{10} + 4\frac{4}{5} \approx$ _____

    c. $9\frac{9}{10} - 5\frac{1}{5} \approx$ _____

    d. $4\frac{1}{9} - 1\frac{1}{10} \approx$ _____

    e. $6\frac{3}{12} + 5\frac{1}{9} \approx$ _____

2. Estimate each sum or difference to the nearest half or whole number by rounding. Explain your estimate using words or a number line.

    a. $\frac{16}{3} + \frac{17}{8} \approx$ _____

    b. $\frac{17}{3} - \frac{15}{4} \approx$ _____

    c. $\frac{57}{8} + \frac{26}{8} \approx$ _____

3. Gina's estimate for $7\frac{5}{8} - 2\frac{1}{2}$ was 5. Dominick's estimate was $5\frac{1}{2}$. Whose estimate do you think is closer to the actual difference? Explain.

4. Use benchmark numbers or mental math to estimate the sum or difference.

a. $10\frac{3}{4} + 12\frac{11}{12}$	b. $2\frac{7}{10} + 23\frac{3}{8}$
c. $15\frac{9}{12} - 8\frac{11}{12}$	d. $\frac{56}{7} - \frac{31}{8}$

EUREKA MATH

1. Solve.

$$6\frac{2}{5} + \frac{3}{5} = 6\frac{5}{5} = 7$$

> I add using unit form. 6 ones 2 fifths + 3 fifths = 6 ones 5 fifths. I know that $\frac{5}{5} = 1$, so $6 + 1 = 7$.

2. Complete the number sentence.

$$18 = 17\frac{3}{10} + \frac{7}{10}$$

> I know that $17 + 1 = 18$, so I need to find a fraction that equals 1 when added to $\frac{3}{10}$. $3 + 7 = 10$, so the fraction that completes the number sentence is 7 tenths.

3. Use a number bond and the arrow way to show how to make one. Solve.

$$3\frac{5}{8} + \frac{6}{8}$$

$$\frac{3}{8} \qquad \frac{3}{8}$$

> I decompose $\frac{6}{8}$ into $\frac{3}{8}$ and $\frac{3}{8}$ because I know $3\frac{5}{8}$ needs $\frac{3}{8}$ to make the next whole number, 4.

$$3\frac{5}{8} \xrightarrow{+\frac{3}{8}} 4 \xrightarrow{+\frac{3}{8}} 4\frac{3}{8}$$

> The arrow way reminds me of making ten or making change from a dollar.

4. Solve.

$$\frac{7}{8} + 4\frac{6}{8}$$

> I can add using any method that makes sense to me, like adding in unit form, using the arrow method, or adding to make the next 1, as shown below.
>
> $$\frac{7}{8} + 4\frac{6}{8} = \frac{5}{8} + 5 = 5\frac{5}{8}$$
>
> $$\frac{5}{8} \qquad \frac{2}{8}$$

$$\frac{7}{8} + 4\frac{6}{8} = 4\frac{13}{8} = 5\frac{5}{8}$$

Name _____   Date _____

1. Solve.

   a. $4\frac{1}{3} + \frac{1}{3}$

   b. $5\frac{1}{4} + \frac{2}{4}$

   c. $\frac{2}{6} + 3\frac{4}{6}$

   d. $\frac{5}{8} + 7\frac{3}{8}$

2. Complete the number sentences.

a. $3\frac{5}{6} +$ _____ $= 4$	b. $5\frac{3}{7} +$ _____ $= 6$
c. $5 = 4\frac{1}{8} +$ _____	d. $15 = 14\frac{4}{12} +$ _____

3. Draw a number bond and the arrow way to show how to make one. Solve.

   a. $2\frac{4}{5} + \frac{2}{5}$ b. $3\frac{2}{3} + \frac{2}{3}$ c. $4\frac{4}{6} + \frac{5}{6}$

   $2\frac{4}{5} \xrightarrow{+\frac{1}{5}} 3 \xrightarrow{+\frac{1}{5}} 3\frac{1}{5}$

4. Solve.

a. $2\frac{3}{5} + \frac{3}{5}$	b. $3\frac{6}{8} + \frac{4}{8}$
c. $5\frac{4}{6} + \frac{3}{6}$	d. $\frac{7}{10} + 6\frac{6}{10}$
e. $\frac{5}{10} + 8\frac{9}{10}$	f. $7\frac{8}{12} + \frac{11}{12}$
g. $3\frac{90}{100} + \frac{58}{100}$	h. $\frac{60}{100} + 14\frac{79}{100}$

Lesson 30: Add a mixed number and a fraction.

EUREKA MATH

5. To solve $4\frac{8}{10} + \frac{3}{10}$, Carmen thought, "$4\frac{8}{10} + \frac{2}{10} = 5$ and $5 + \frac{1}{10} = 5\frac{1}{10}$."

Benny thought, "$4\frac{8}{10} + \frac{3}{10} = 4\frac{11}{10} = 4 + \frac{10}{10} + \frac{1}{10} = 5\frac{1}{10}$." Explain why Carmen and Benny are both right.

1. Solve.

$$3\frac{1}{5} + 2\frac{4}{5}$$

> I can add like units. 3 ones 1 fifth + 2 ones 4 fifths = 5 ones 5 fifths.

$$3\frac{1}{5} + 2\frac{4}{5} = 5 + \frac{5}{5} = 5 + 1 = 6$$

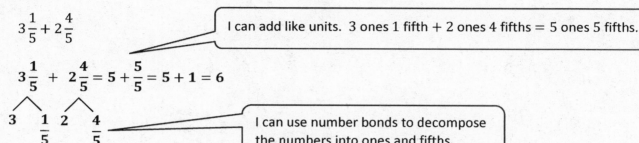

3   $\frac{1}{5}$ 2 $\frac{4}{5}$

> I can use number bonds to decompose the numbers into ones and fifths.

2. Solve. Use a number line to show your work.

$$1\frac{2}{3} + 3\frac{2}{3}$$

$$1\frac{2}{3} + 3\frac{2}{3} = 4 + \frac{4}{3} = 5\frac{1}{3}$$

$\frac{3}{3}$ $\frac{1}{3}$

$+\frac{3}{3}$ $+\frac{1}{3}$

2   3   4   5 $5\frac{1}{3}$ 6

> I add the ones and thirds. I decompose $\frac{4}{3}$ into 1 and $\frac{1}{3}$. $4 + 1 + \frac{1}{3} = 5\frac{1}{3}$

3. Solve. Use the arrow way to show how to make one.

$$4\frac{7}{12} + 3\frac{9}{12}$$

$$4\frac{7}{12} + 3\frac{9}{12} = 7\frac{7}{12} + \frac{9}{12} = 8\frac{4}{12}$$

$\frac{5}{12}$ $\frac{4}{12}$

$7\frac{7}{12} \xrightarrow{+\frac{5}{12}} 8 \xrightarrow{+\frac{4}{12}} 8\frac{4}{12}$

> I use the arrow way to add $\frac{5}{12}$ and $7\frac{7}{12}$ to make the next whole number. Then, I add the other part of the number bond to get $8\frac{4}{12}$.

Name _____   Date _____

1. Solve.

   a. $2\frac{1}{3}$ + $1\frac{2}{3}$ = $3 + \frac{3}{3}$ =

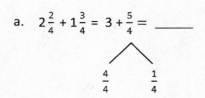

   2 $\frac{1}{3}$ 1 $\frac{2}{3}$

   b. $2\frac{2}{5} + 2\frac{2}{5}$

   c. $3\frac{3}{8} + 1\frac{5}{8}$

2. Solve. Use a number line to show your work.

   a. $2\frac{2}{4} + 1\frac{3}{4}$ = $3 + \frac{5}{4}$ = _____

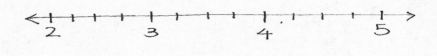

      $\frac{4}{4}$ $\frac{1}{4}$

   b. $3\frac{4}{6} + 2\frac{5}{6}$

   c. $1\frac{9}{12} + 1\frac{7}{12}$

3. Solve. Use the arrow way to show how to make one.

    a. $2\frac{3}{4} + 1\frac{3}{4} = 3\frac{3}{4} + \frac{3}{4} =$

    $\frac{1}{4}$ $\frac{2}{4}$

    $3\frac{3}{4} \xrightarrow{+\frac{1}{4}} 4 \longrightarrow$

    b. $2\frac{7}{8} + 3\frac{4}{8}$

    c. $1\frac{7}{9} + 4\frac{5}{9}$

4. Solve. Use whichever method you prefer.

    a. $1\frac{4}{5} + 1\frac{3}{5}$

    b. $3\frac{8}{10} + 1\frac{5}{10}$

    c. $2\frac{5}{7} + 3\frac{6}{7}$

Lesson 31:   Add mixed numbers.

EUREKA MATH

1. Subtract. Model with a number line or the arrow way.

$4\frac{3}{5} - \frac{2}{5} = 4\frac{1}{5}$

> I can subtract 2 fifths $\frac{1}{5}$ at a time or all at once.

$4\frac{3}{5} \xrightarrow{-\frac{1}{5}} 4\frac{2}{5} \xrightarrow{-\frac{1}{5}} 4\frac{1}{5}$

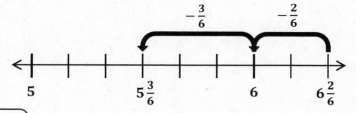

2. Use decomposition to subtract the fractions. Model with a number line or the arrow way.

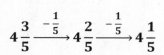

> I decompose $\frac{5}{6}$ into $\frac{2}{6}$ and $\frac{3}{6}$ so that I can subtract $\frac{2}{6}$ from $6\frac{2}{6}$ to get to a whole number.

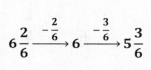

$6\frac{2}{6} \xrightarrow{-\frac{2}{6}} 6 \xrightarrow{-\frac{3}{6}} 5\frac{3}{6}$

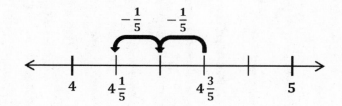

> I subtract the other part of the number bond, $\frac{3}{6}$.

3. Decompose the total to subtract the fraction.

$8\frac{2}{12} - \frac{9}{12}$

> There aren't enough twelfths to subtract 9 twelfths, so I decompose the total to subtract $\frac{9}{12}$ from 1.

$8\frac{2}{12} - \frac{9}{12} = 7\frac{2}{12} + \frac{3}{12} = 7\frac{5}{12}$

$7\frac{2}{12} \quad 1$

> Once $\frac{9}{12}$ is subtracted, the remaining numbers are added together.

Name _____   Date _____

1. Subtract. Model with a number line or the arrow way.

    a. $6\frac{3}{5} - \frac{1}{5}$

    b. $4\frac{9}{12} - \frac{7}{12}$

    c. $7\frac{1}{4} - \frac{3}{4}$

    d. $8\frac{3}{8} - \frac{5}{8}$

2. Use decomposition to subtract the fractions. Model with a number line or the arrow way.

    a. $2\frac{2}{5} - \frac{4}{5}$

    $\frac{2}{5}$ $\frac{2}{5}$

    b. $2\frac{1}{3} - \frac{2}{3}$

    c. $4\frac{1}{6} - \frac{4}{6}$

    d. $3\frac{3}{6} - \frac{5}{6}$

EUREKA
MATH

e. $9\frac{3}{8} - \frac{7}{8}$

f. $7\frac{1}{10} - \frac{6}{10}$

g. $10\frac{1}{8} - \frac{5}{8}$

h. $9\frac{4}{12} - \frac{7}{12}$

i. $11\frac{3}{5} - \frac{4}{5}$

j. $17\frac{1}{9} - \frac{5}{9}$

3. Decompose the total to subtract the fractions.

a. $4\frac{1}{8} - \frac{3}{8} = 3\frac{1}{8} + \frac{5}{8} = 3\frac{6}{8}$

$3\frac{1}{8}$ ⌃ 1

b. $5\frac{2}{5} - \frac{3}{5}$

c. $7\frac{1}{8} - \frac{3}{8}$

d. $3\frac{3}{9} - \frac{4}{9}$

e. $6\frac{3}{10} - \frac{7}{10}$

f. $2\frac{5}{9} - \frac{8}{9}$

Lesson 32:   Subtract a fraction from a mixed number.

EUREKA MATH®

1. Write a related addition sentence. Subtract by counting on. Use a number line or the arrow way to help.

$$6\frac{1}{4} - 2\frac{3}{4} = 3\frac{2}{4}$$

I add the numbers on top of the arrows to find the unknown addend.

$$\frac{1}{4} + 3 + \frac{1}{4} = 3\frac{2}{4}$$

$$2\frac{3}{4} + 3\frac{2}{4} = 6\frac{1}{4}$$

$$2\frac{3}{4} \xrightarrow{+\frac{1}{4}} 3 \xrightarrow{+3} 6 \xrightarrow{+\frac{1}{4}} 6\frac{1}{4}$$

I add 3 to get to 6.

I use the arrow way to count up to solve for the unknown in my addition sentence. I add $\frac{1}{4}$ to get to the next one, 3.

My final number needs to be $6\frac{1}{4}$, so I need to add 1 more fourth.

2. Subtract by decomposing the fractional part of the number you are subtracting. Use a number line or the arrow way to help you.

$$4\frac{1}{3} - 1\frac{2}{3} = 3\frac{1}{3} - \frac{2}{3} = 2\frac{2}{3}$$

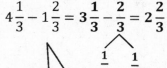

$$\frac{1}{3} \quad \frac{1}{3}$$

I subtract 1 from $4\frac{1}{3}$.

$3\frac{1}{3} - \frac{1}{3} = 3$ and $3 - \frac{1}{3} = 2\frac{2}{3}$.

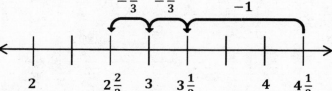

$$-\frac{1}{3} \quad -\frac{1}{3} \qquad -1$$

$$2 \qquad\qquad 2\frac{2}{3} \quad 3 \quad 3\frac{1}{3} \qquad 4 \quad 4\frac{1}{3}$$

EUREKA MATH®

3. Subtract by decomposing to take one out.

$$7\frac{2}{10} - 5\frac{9}{10}$$

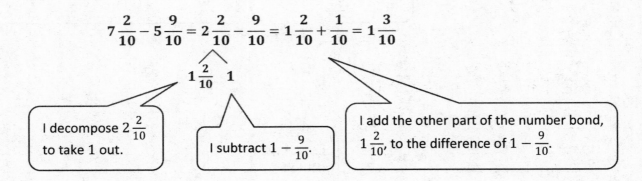

$$7\frac{2}{10} - 5\frac{9}{10} = 2\frac{2}{10} - \frac{9}{10} = 1\frac{2}{10} + \frac{1}{10} = 1\frac{3}{10}$$

$1\frac{2}{10}$ 1

I decompose $2\frac{2}{10}$ to take 1 out.

I subtract $1 - \frac{9}{10}$.

I add the other part of the number bond, $1\frac{2}{10}$, to the difference of $1 - \frac{9}{10}$.

Lesson 33:   Subtract a mixed number from a mixed number.

EUREKA MATH®

Name _____ Date _____

1. Write a related addition sentence. Subtract by counting on. Use a number line or the arrow way to help. The first one has been partially done for you.

   a. $3\frac{2}{5} - 1\frac{4}{5} =$ _____

   $1\frac{4}{5} +$ _____ $= 3\frac{2}{5}$

   b. $5\frac{3}{8} - 2\frac{5}{8}$

2. Subtract, as shown in Problem 2(a) below, by decomposing the fractional part of the number you are subtracting. Use a number line or the arrow way to help you.

   a. $4\frac{1}{5} - 1\frac{3}{5} = 3\frac{1}{5} - \frac{3}{5} = 2\frac{3}{5}$

   $\overbrace{}$
   $\frac{1}{5} \qquad \frac{2}{5}$

   b. $4\frac{1}{7} - 2\frac{4}{7}$

   c. $5\frac{5}{12} - 3\frac{8}{12}$

3. Subtract, as shown in 3(a) below, by decomposing to take one out.

a. $5\frac{5}{8} - 2\frac{7}{8} = 3\frac{5}{8} - \frac{7}{8} =$

$2\frac{5}{8}$ 1

b. $4\frac{3}{12} - 3\frac{8}{12}$

c. $9\frac{1}{10} - 6\frac{9}{10}$

4. Solve using any strategy.

a. $6\frac{1}{9} - 4\frac{3}{9}$ b. $5\frac{3}{10} - 3\frac{6}{10}$

c. $8\frac{7}{12} - 5\frac{9}{12}$ d. $7\frac{4}{100} - 2\frac{92}{100}$

Lesson 33:   Subtract a mixed number from a mixed number.   EUREKA
                                                                                          MATH

1. Subtract.

$$8\frac{2}{7} - \frac{6}{7} = 7\frac{9}{7} - \frac{6}{7} = 7\frac{3}{7}$$

7    $\frac{9}{7}$

Now I have 9 sevenths, which is enough sevenths to subtract 6 sevenths.

It's just like renaming 1 ten for 10 ones when subtracting whole numbers, except I rename 1 one for 7 sevenths.

2. Subtract the ones first.

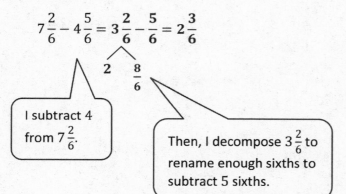

$$7\frac{2}{6} - 4\frac{5}{6} = 3\frac{2}{6} - \frac{5}{6} = 2\frac{3}{6}$$

2    $\frac{8}{6}$

I subtract 4 from $7\frac{2}{6}$.

Then, I decompose $3\frac{2}{6}$ to rename enough sixths to subtract 5 sixths.

$$7\frac{2}{6} \xrightarrow{-4} 3\frac{2}{6} \xrightarrow{-\frac{5}{6}} 2\frac{3}{6}$$

I can show the same work with the arrow way.

Name _____   Date _____

1. Subtract.

a. $5\frac{1}{4} - \frac{3}{4}$

4          $\frac{5}{4}$

b. $6\frac{3}{8} - \frac{6}{8}$

c. $7\frac{4}{6} - \frac{5}{6}$

2. Subtract the ones first.

a. $4\frac{1}{5} - 1\frac{3}{5} = 3\frac{1}{5} - \frac{3}{5} = 2\frac{3}{5}$

2          $\frac{6}{5}$

b. $4\frac{3}{6} - 2\frac{5}{6}$

EUREKA
MATH

c. $8\frac{3}{8} - 2\frac{5}{8}$

d. $13\frac{3}{10} - 8\frac{7}{10}$

3. Solve using any strategy.

a. $7\frac{3}{12} - 4\frac{9}{12}$

b. $9\frac{6}{10} - 5\frac{8}{10}$

c. $17\frac{2}{16} - 9\frac{7}{16}$

d. $12\frac{5}{100} - 8\frac{94}{100}$

Lesson 34:   Subtract mixed numbers.

EUREKA
MATH®

1. Draw and label a tape diagram to show the following is true:

> I can move the parentheses in the equation, associating the factors, 5 and 2. When I do so, fourths becomes the unit.

10 fourths = 5 × (2 fourths) = (5 × 2) fourths

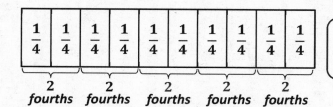

> I can do this with any unit:
> 10 bananas = 5 × (2 bananas) = (5 × 2) bananas.

2   2   2   2   2
fourths fourths fourths fourths fourths

> Using brackets to group every 2 units of $\frac{1}{4}$, I model 5 copies of 2 fourths.

> The product of 5 and 2 is 10. My model shows that (5 × 2) fourths is the same as 5 × (2 fourths), or 10 fourths.

2. Write the equation in unit form to solve.

$$8 \times \frac{2}{3} = \frac{16}{3}$$

8 × 2 thirds = 16 thirds

> Unit form simplifies my multiplication. Instead of puzzling over how to multiply a fraction by a whole number, I unveil an easy fact I can solve fast! I know 8 × 2 is 16, so 8 × 2 thirds is 16 thirds.

3. Solve.

$$6 \times \frac{3}{4}$$

> The unit is fourths! I think in unit form, 6 × 3 fourths is 18 fourths.

$$6 \times \frac{3}{4} = \frac{6 \times 3}{4} = \frac{18}{4}$$

Lesson 35: Represent the multiplication of *n* times *a/b* as (*n × a*)/*b* using the associative property and visual models.

© 2018 Great Minds®. eureka-math.org

EUREKA MATH

155

4. Ms. Swanson bought some apple juice. Each member of her family drank $\frac{3}{5}$ cup for breakfast. Including Ms. Swanson, there are four people in her family. How many cups of apple juice did they drink?

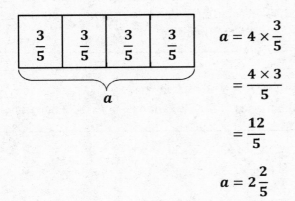

$$a = 4 \times \frac{3}{5}$$

$$= \frac{4 \times 3}{5}$$

$$= \frac{12}{5}$$

$$a = 2\frac{2}{5}$$

Ms. Swanson and her family drank $2\frac{2}{5}$ cups of apple juice.

Lesson 35:   Represent the multiplication of *n* times *a/b* as (*n* × *a*)/*b* using the associative property and visual models.

EUREKA
MATH®

Name _____ Date _____

1. Draw and label a tape diagram to show the following are true.

   a. 8 thirds = 4 × (2 thirds) = (4 × 2) thirds

   b. 15 eighths = 3 × (5 eighths) = (3 × 5) eighths

2. Write the expression in unit form to solve.

   a. $10 \times \frac{2}{5}$

   b. $3 \times \frac{5}{6}$

   c. $9 \times \frac{4}{9}$

   d. $7 \times \frac{3}{4}$

EUREKA MATH

Lesson 35: Represent the multiplication of n times a/b as (n × a)/b using the associative property and visual models.

© 2018 Great Minds®. eureka-math.org

157

3. Solve.

   a. $6 \times \frac{3}{4}$

   b. $7 \times \frac{5}{8}$

   c. $13 \times \frac{2}{3}$

   d. $18 \times \frac{2}{3}$

   e. $14 \times \frac{7}{10}$

   f. $7 \times \frac{14}{100}$

4. Mrs. Smith bought some orange juice. Each member of her family drank $\frac{2}{3}$ cup for breakfast. There are five people in her family. How many cups of orange juice did they drink?

Lesson 35: Represent the multiplication of *n* times *a/b* as (*n* × *a*)/*b* using the associative property and visual models.

EUREKA MATH

1. Draw a tape diagram to represent $\frac{3}{8} + \frac{3}{8} + \frac{3}{8} + \frac{3}{8}$.

| $\frac{3}{8}$ | $\frac{3}{8}$ | $\frac{3}{8}$ | $\frac{3}{8}$ |

> I model 4 copies of $\frac{3}{8}$.

Write a multiplication expression equal to $\frac{3}{8} + \frac{3}{8} + \frac{3}{8} + \frac{3}{8}$.

$$4 \times \frac{3}{8} = \frac{12}{8} = 1\frac{4}{8} = 1\frac{1}{2}$$

> Multiplication is more efficient than addition. I can solve easily by thinking in unit form: 4×3 eighths is 12 eighths.

2. Solve using any method. Express your answers as whole or mixed numbers.

a. $4 \times \frac{5}{8}$

| $\frac{5}{8}$ | $\frac{5}{8}$ | $\frac{5}{8}$ | $\frac{5}{8}$ |

$$4 \times \frac{5}{8} = \frac{4 \times 5}{8} = \frac{20}{8} = 2\frac{4}{8} = 2\frac{1}{2}$$

b. $32 \times \frac{2}{5}$

$$32 \times \frac{2}{5} = 32 \times 2 \textit{ fifths} = 64 \textit{ fifths} = \frac{64}{5} = 12\frac{4}{5}$$

> To solve, I think to myself, 5 times what number is close to or equal to 64? Or, I can divide 64 by 5.

3. A bricklayer places 13 bricks end to end along the entire outside length of a shed's wall. Each brick is $\frac{2}{3}$ foot long. How long is that wall of the shed?

| $\frac{2}{3}$ | $\frac{2}{3}$ | $\frac{2}{3}$ | $\frac{2}{3}$ | $\frac{2}{3}$ | $\frac{2}{3}$ | $\frac{2}{3}$ | $\frac{2}{3}$ | $\frac{2}{3}$ | $\frac{2}{3}$ | $\frac{2}{3}$ | $\frac{2}{3}$ | $\frac{2}{3}$ |

s

$$13 \times \frac{2}{3} = \frac{13 \times 2}{3} = \frac{26}{3} = 8\frac{2}{3}$$

The wall of the shed is $8\frac{2}{3}$ feet long.

> It would take too long to write an addition sentence to solve! Multiplication is quick and easy!

EUREKA MATH

Lesson 36: Represent the multiplication of n times a/b as $(n \times a)/b$ using the associative property and visual models.

159

© 2018 Great Minds®. eureka-math.org

Name _____   Date _____

1. Draw a tape diagram to represent
   $\frac{2}{3} + \frac{2}{3} + \frac{2}{3} + \frac{2}{3}$.

2. Draw a tape diagram to represent
   $\frac{7}{8} + \frac{7}{8} + \frac{7}{8}$.

   Write a multiplication expression equal to
   $\frac{2}{3} + \frac{2}{3} + \frac{2}{3} + \frac{2}{3}$.

   Write a multiplication expression equal to
   $\frac{7}{8} + \frac{7}{8} + \frac{7}{8}$.

3. Rewrite each repeated addition problem as a multiplication problem and solve. Express the result as a mixed number. The first one has been completed for you.

   a. $\frac{7}{5} + \frac{7}{5} + \frac{7}{5} + \frac{7}{5} = 4 \times \frac{7}{5} = \frac{4 \times 7}{5} = \frac{28}{5} = 5\frac{3}{5}$

   b. $\frac{7}{10} + \frac{7}{10} + \frac{7}{10}$

   c. $\frac{5}{12} + \frac{5}{12} + \frac{5}{12} + \frac{5}{12} + \frac{5}{12} + \frac{5}{12}$

   d. $\frac{3}{8} + \frac{3}{8} + \frac{3}{8} + \frac{3}{8} + \frac{3}{8} + \frac{3}{8} + \frac{3}{8} + \frac{3}{8} + \frac{3}{8} + \frac{3}{8} + \frac{3}{8} + \frac{3}{8}$

4. Solve using any method. Express your answers as whole or mixed numbers.

   a. $7 \times \frac{2}{9}$

   b. $11 \times \frac{2}{3}$

EUREKA MATH

Lesson 36:   Represent the multiplication of *n* times *a/b* as (*n* × *a*)/*b* using the associative property and visual models.

© 2018 Great Minds®. eureka-math.org

161

c. $40 \times \frac{2}{6}$

d. $24 \times \frac{5}{6}$

e. $23 \times \frac{3}{5}$

f. $34 \times \frac{2}{8}$

5. Coleton is playing with interlocking blocks that are each $\frac{3}{4}$ inch tall. He makes a tower 17 blocks tall. How tall is his tower in inches?

6. There were 11 players on Mr. Maiorani's softball team. They each ate $\frac{3}{8}$ of a pizza. How many pizzas did they eat?

7. A bricklayer places 12 bricks end to end along the entire outside length of a shed's wall. Each brick is $\frac{3}{4}$ foot long. How many feet long is that wall of the shed?

Lesson 36:   Represent the multiplication of *n* times *a/b* as (*n* × *a*)/*b* using the associative property and visual models.

© 2018 Great Minds®. eureka-math.org

EUREKA
MATH

1. Draw tape diagrams to show two ways to represent 3 units of $5\frac{1}{12}$.

$5\frac{1}{12}$	$5\frac{1}{12}$	$5\frac{1}{12}$

5	5	5	$\frac{1}{12}$ $\frac{1}{12}$ $\frac{1}{12}$

> I rearrange the model for 3 copies of $5\frac{1}{12}$ by decomposing $5\frac{1}{12}$ into two parts: 5 and $\frac{1}{12}$. I show 3 groups of 5 and 3 groups of $\frac{1}{12}$.

Write a multiplication expression to match each tape diagram.

$$3 \times 5\frac{1}{12}$$

$$(3 \times 5) + \left(3 \times \frac{1}{12}\right)$$

> $5\frac{1}{12}$ is composed of two units: ones and twelfths. I use the distributive property to multiply the value of each unit by 3. $3 \times 5\frac{1}{12}$ is equal to 3 fives and 3 twelfths.

2. Solve using the distributive property.

a. $2 \times 3\frac{5}{6} = 2 \times \left(3 + \frac{5}{6}\right)$

$$= (2 \times 3) + \left(2 \times \frac{5}{6}\right)$$
$$= 6 + \frac{10}{6}$$
$$= 6 + 1\frac{4}{6}$$
$$= 7\frac{4}{6}$$

> I omit writing this step for Part (b) because I can see it's 4 copies of 2 and 4 copies of $\frac{3}{4}$, or $8 + \frac{12}{4}$.

b. $4 \times 2\frac{3}{4} = 4 \times \left(2 + \frac{3}{4}\right)$

$$= 8 + \frac{12}{4}$$
$$= 8 + 3$$
$$= 11$$

3. Sara's street is $1\frac{3}{5}$ miles long. She ran the length of the street 3 times. How far did she run?

| $1\frac{3}{5}$ | $1\frac{3}{5}$ | $1\frac{3}{5}$ |

s

$$s = 3 \times 1\frac{3}{5}$$

$$= (3 \times 1) + \left(3 \times \frac{3}{5}\right)$$

I use the distributive property to multiply the ones by 3 and the fractional part by 3.

$$= 3 + \frac{9}{5}$$

$$= 3 + 1\frac{4}{5}$$

$$s = 4\frac{4}{5}$$

Sara ran $4\frac{4}{5}$ miles.

Lesson 37:   Find the product of a whole number and a mixed number using the distributive property.

EUREKA MATH

Name _____ Date _____

1. Draw tape diagrams to show two ways to represent 3 units of $5\frac{1}{12}$.

   Write a multiplication expression to match each tape diagram.

2. Solve the following using the distributive property. The first one has been done for you. (As soon as you are ready, you may omit the step that is in line 2.)

a. $3 \times 6\frac{4}{5} = 3 \times \left(6 + \frac{4}{5}\right)$ $\qquad = (3 \times 6) + \left(3 + \frac{4}{5}\right)$ $\qquad = 18 + \frac{12}{5}$ $\qquad = 18 + 2\frac{2}{5}$ $\qquad = 20\frac{2}{5}$	b. $5 \times 4\frac{1}{6}$
c. $6 \times 2\frac{3}{5}$	d. $2 \times 7\frac{3}{10}$

EUREKA MATH

Lesson 37:   Find the product of a whole number and a mixed number using the distributive property.

165

© 2018 Great Minds®. eureka-math.org

e. $8 \times 7\frac{1}{4}$	f. $3\frac{3}{8} \times 12$

3. Sara's street is $2\frac{3}{10}$ miles long. She ran the length of the street 6 times. How far did she run?

4. Kelly's new puppy weighed $4\frac{7}{10}$ pounds when she brought him home. Now, he weighs six times as much. How much does he weigh now?

Lesson 37:   Find the product of a whole number and a mixed number using the distributive property.

EUREKA MATH

1. Fill in the unknown factors.

   a. $7 \times 3\frac{4}{5} = (\underline{\ \ 7\ \ } \times 3) + (\underline{\ \ 7\ \ } \times \frac{4}{5})$

   b. $6 \times 4\frac{3}{8} = (6 \times \underline{\ \ 4\ \ }) + (6 \times \underline{\ \ \frac{3}{8}\ \ })$

   > The mixed number is distributed as the whole and the fraction. Both of the distributed numbers have to be multiplied by 7, so 7 is the missing factor.

2. Multiply. Use the distributive property.

   $5 \times 7\frac{3}{5}$

7	$\frac{3}{5}$	7	$\frac{3}{5}$	7	$\frac{3}{5}$	7	$\frac{3}{5}$	7	$\frac{3}{5}$

   $$5 \times 7\frac{3}{5} = 35 + \frac{15}{5}$$
   $$= 35 + 3$$
   $$= 38$$

   > I break apart $7\frac{3}{5}$ into 7 and $\frac{3}{5}$. 5 sevens equals 35, and 5 copies of 3 fifths equals 15 fifths, or 3.

3. Amina's dog ate $2\frac{2}{3}$ cups of dog food each day for three weeks. How much dog food did Amina's dog eat during the three weeks?

   > There are 7 days in a week. To find the number of days in 3 weeks, I multiply 7×3. There are 21 days in 3 weeks.

   $$21 \times 2\frac{2}{3} = 42 + \frac{42}{3}$$
   $$= 42 + 14$$
   $$= 56$$

   Amina's dog ate 56 cups of food during the three weeks.

EUREKA MATH

Lesson 38: Find the product of a whole number and a mixed number using the distributive property.

167

© 2018 Great Minds®. eureka-math.org

Name _____ Date _____

1. Fill in the unknown factors.

    a. $8 \times 4\frac{4}{7} = (\underline{} \times 4) + (\underline{} \times \frac{4}{7})$ b. $9 \times 7\frac{7}{10} = (9 \times \underline{}) + (9 \times \underline{})$

2. Multiply. Use the distributive property.

    a. $6 \times 8\frac{2}{7}$

    b. $7\frac{3}{4} \times 9$

    c. $9 \times 8\frac{7}{9}$

    d. $25\frac{7}{8} \times 3$

EUREKA MATH   **Lesson 38:**   Find the product of a whole number and a mixed number using the distributive property.   **169**

© 2018 Great Minds®. eureka-math.org

e. $4 \times 20\frac{8}{12}$

f. $30\frac{3}{100} \times 12$

3. Brandon is cutting 9 boards for a woodworking project. Each board is $4\frac{5}{8}$ feet long. What is the total length of the boards?

4. Rocky the collie ate $3\frac{1}{4}$ cups of dog food each day for two weeks. How much dog food did Rocky eat in that time?

5. At the class party, each student will be given a container filled with $8\frac{5}{8}$ ounces of juice. There are 25 students in the class. How many ounces of juice does the teacher need to buy?

Lesson 38:   Find the product of a whole number and a mixed number using the distributive property.

EUREKA
MATH

1. It takes $9\frac{2}{3}$ yards of yarn to make one baby blanket. Upik needs four times as much yarn to make four baby blankets. She already has 6 yards of yarn. How many more yards of yarn does Upik need to buy in order to make four baby blankets?

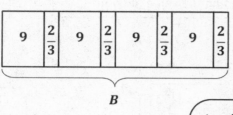

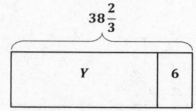

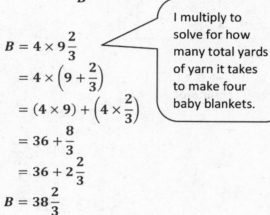

I multiply to solve for how many total yards of yarn it takes to make four baby blankets.

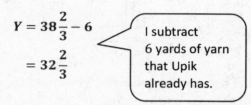

$$Y = 38\frac{2}{3} - 6$$

$$= 32\frac{2}{3}$$

I subtract 6 yards of yarn that Upik already has.

Upik needs to buy $32\frac{2}{3}$ more yards of yarn.

EUREKA
MATH®

2. The caterpillar crawled $34\frac{2}{3}$ centimeters on Monday. He crawled 5 times as far on Tuesday. How far did he crawl in the two days?

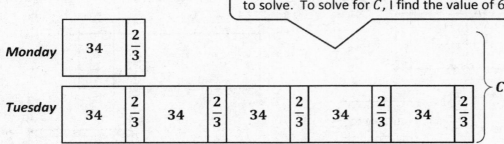

I use the tape diagram to find the most efficient way to solve. To solve for C, I find the value of 6 units.

The caterpillar crawled 208 centimeters, or 2 meters 8 centimeters, on Monday and Tuesday.

$$C = 6 \times 34\frac{2}{3}$$

$$C = (6 \times 34) + \left(6 \times \frac{2}{3}\right)$$

$$C = 204 + \frac{12}{3}$$

$$C = 204 + 4$$

$$C = 208$$

EUREKA
MATH®

Name _____ Date _____

Use the RDW process to solve.

1. Ground turkey is sold in packages of $2\frac{1}{2}$ pounds. Dawn bought eight times as much turkey that is sold in 1 package for her son's birthday party. How many pounds of ground turkey did Dawn buy?

2. Trevor's stack of books is $7\frac{7}{4}$ inches tall. Rick's stack is 3 times as tall. What is the difference in the heights of their stacks of books?

3. It takes $8\frac{3}{4}$ yards of fabric to make one quilt. Gail needs three times as much fabric to make three quilts. She already has two yards of fabric. How many more yards of fabric does Gail need to buy in order to make three quilts?

4. Carol made punch. She used $12\frac{3}{8}$ cups of juice and then added three times as much ginger ale. Then, she added 1 cup of lemonade. How many cups of punch did her recipe make?

5. Brandon drove $72\frac{7}{10}$ miles on Monday. He drove 3 times as far on Tuesday. How far did he drive in the two days?

6. Mrs. Reiser used $9\frac{8}{10}$ gallons of gas this week. Mr. Reiser used five times as much gas as Mrs. Reiser used this week. If Mr. Reiser pays $3 for each gallon of gas, how much did Mr. Reiser pay for gas this week?

Lesson 39: Solve multiplicative comparison word problems involving fractions.

EUREKA MATH

Noura recorded the growth of her plant during the year.

The measurements are listed in the table.

1. Use the data to create a line plot.

> I remember making a line plot in Lesson 28.

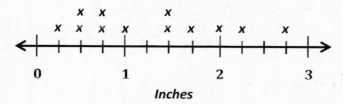

Growth of Plant

x = growth in one month

Month	Growth of Plant (in inches)
January	$\frac{1}{2}$
February	$\frac{3}{4}$
March	$1\frac{1}{2}$
April	2
May	$1\frac{1}{4}$
June	$1\frac{3}{4}$
July	$2\frac{3}{4}$
August	$2\frac{1}{4}$
September	1
October	$\frac{3}{4}$
November	$\frac{1}{2}$
December	$\frac{1}{4}$

2. How many inches did Noura's plant grow in the spring months of March, April, and May?

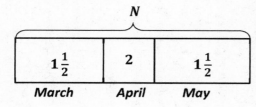

> I add the whole numbers first!

$$N = 1\frac{1}{2} + 1 + 1\frac{1}{2}$$
$$N = 3 + \frac{2}{2}$$
$$N = 4$$

Noura's plant grew a total of 4 inches during the spring months.

3. In which months did her plant grow twice as many inches as it did in October?

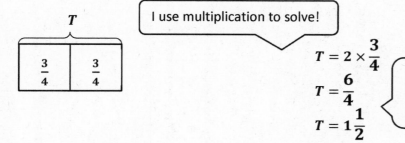

> I use multiplication to solve!

$$T = 2 \times \frac{3}{4}$$
$$T = \frac{6}{4}$$
$$T = 1\frac{1}{2}$$

> I can use a number bond or number line to help rename a fraction to a mixed number, if needed.

Noura's plant grew twice as many inches in the months of May and March as it did in October.

Lesson 40: Solve word problems involving the multiplication of a whole number and a fraction including those involving line plots.

175

EUREKA MATH®

Name _____   Date _____

The chart to the right shows the total monthly rainfall for a city.

1. Use the data to create a line plot at the bottom of this page and to answer the following questions.

Month	Rainfall (in inches)
January	$2\frac{2}{8}$
February	$1\frac{3}{8}$
March	$2\frac{3}{8}$
April	$2\frac{5}{8}$
May	$4\frac{1}{4}$
June	$2\frac{1}{4}$
July	$3\frac{7}{8}$
August	$3\frac{1}{4}$
September	$1\frac{5}{8}$
October	$3\frac{2}{8}$
November	$1\frac{3}{4}$
December	$1\frac{5}{8}$

Lesson 40: Solve word problems involving the multiplication of a whole number and a fraction including those involving line plots.

177

EUREKA MATH®

2. What is the difference in rainfall from the wettest and driest months?

3. How much more rain fell in May than in April?

4. What is the combined rainfall amount for the summer months of June, July, and August?

5. How much more rain fell in the summer months than the combined rainfall for the last 4 months of the year?

6. In which months did it rain twice as much as it rained in December?

7. Each inch of rain can produce ten times that many inches of snow. If all of the rainfall in January was in the form of snow, how many inches of snow fell in January?

Lesson 40:   Solve word problems involving the multiplication of a whole number and a fraction including those involving line plots.

EUREKA MATH

1. Find the sums.

> I draw brackets connecting fractions that add up to equal 1.

> There are 2 pairs of fractions that equal 1. 2 fourths is leftover without a partner.

a. $\dfrac{0}{3} + \dfrac{1}{3} + \dfrac{2}{3} + \dfrac{3}{3}$

b. $\dfrac{0}{4} + \dfrac{1}{4} + \dfrac{2}{4} + \dfrac{3}{4} + \dfrac{4}{4}$

$$\left(\dfrac{0}{3} + \dfrac{3}{3}\right) + \left(\dfrac{1}{3} + \dfrac{2}{3}\right) = 1 + 1 = 2$$

$$\left(\dfrac{0}{4} + \dfrac{4}{4}\right) + \left(\dfrac{1}{4} + \dfrac{3}{4}\right) + \dfrac{2}{4} = 1 + 1 + \dfrac{1}{2} = 2\dfrac{1}{2}$$

> The denominator is odd. Every addend has a partner.

> The denominator is even. One addend does not have a partner. This could be a pattern.

2. Find the sums.

> I notice patterns that help me solve without calculating!

a. $\dfrac{0}{13} + \dfrac{1}{13} + \dfrac{2}{13} + \cdots + \dfrac{13}{13}$

7

b. $\dfrac{0}{16} + \dfrac{1}{16} + \dfrac{2}{16} + \cdots + \dfrac{16}{16}$

$8\dfrac{8}{16}$

> I think about the number of addends, 14, in the expression with odd denominators.

> There are 17 addends in this expression with even denominators. Half of 17 is $8\dfrac{1}{2}$.

3. How can you apply this strategy to find the sum of all the whole numbers from 0 to 1,000?

Sample Student Response:

I can pair the 1,001 addends from 0 to 1,000 to make sums that equal 1,000. There would be 500 pairs. One addend would be left over. I multiply 1,000 × 500, which makes 500,000. When I add the left over addend, I have a total sum of 500,500.

Lesson 41: Find and use a pattern to calculate the sum of all fractional parts between 0 and 1. Share and critique peer strategies.

179

© 2018 Great Minds®. eureka-math.org

Name _____ Date _____

1. Find the sums.

   a. $\frac{0}{5} + \frac{1}{5} + \frac{2}{5} + \frac{3}{5} + \frac{4}{5} + \frac{5}{5}$

   b. $\frac{0}{6} + \frac{1}{6} + \frac{2}{6} + \frac{3}{6} + \frac{4}{6} + \frac{5}{6} + \frac{6}{6}$

   c. $\frac{0}{7} + \frac{1}{7} + \frac{2}{7} + \frac{3}{7} + \frac{4}{7} + \frac{5}{7} + \frac{6}{7} + \frac{7}{7}$

   d. $\frac{0}{8} + \frac{1}{8} + \frac{2}{8} + \frac{3}{8} + \frac{4}{8} + \frac{5}{8} + \frac{6}{8} + \frac{7}{8} + \frac{8}{8}$

   e. $\frac{0}{9} + \frac{1}{9} + \frac{2}{9} + \frac{3}{9} + \frac{4}{9} + \frac{5}{9} + \frac{6}{9} + \frac{7}{9} + \frac{8}{9} + \frac{9}{9}$

   f. $\frac{0}{10} + \frac{1}{10} + \frac{2}{10} + \frac{3}{10} + \frac{4}{10} + \frac{5}{10} + \frac{6}{10} + \frac{7}{10} + \frac{8}{10} + \frac{9}{10} + \frac{10}{10}$

2. Describe a pattern you notice when adding the sums of fractions with even denominators as opposed to those with odd denominators.

3. How would the sums change if the addition started with the unit fraction rather than with 0?

EUREKA MATH®

Lesson 41:   Find and use a pattern to calculate the sum of all fractional parts between 0 and 1. Share and critique peer strategies.

181

© 2018 Great Minds®. eureka-math.org

4. Find the sums.

a. $\frac{0}{20} + \frac{1}{20} + \frac{2}{20} + \cdots + \frac{20}{20}$

b. $\frac{0}{35} + \frac{1}{35} + \frac{2}{35} + \cdots + \frac{35}{35}$

c. $\frac{0}{36} + \frac{1}{36} + \frac{2}{36} + \cdots + \frac{36}{36}$

d. $\frac{0}{75} + \frac{1}{75} + \frac{2}{75} + \cdots + \frac{75}{75}$

e. $\frac{0}{100} + \frac{1}{100} + \frac{2}{100} + \cdots + \frac{100}{100}$

f. $\frac{0}{99} + \frac{1}{99} + \frac{2}{99} + \cdots + \frac{99}{99}$

5. How can you apply this strategy to find the sum of all the whole numbers from 0 to 50? To 99?

Lesson 41: Find and use a pattern to calculate the sum of all fractional parts
between 0 and 1. Share and critique peer strategies.

EUREKA
MATH

Grade 4
Module 6

1. Shade the bottle to show the correct amount. Write the total amount of water in fraction form.

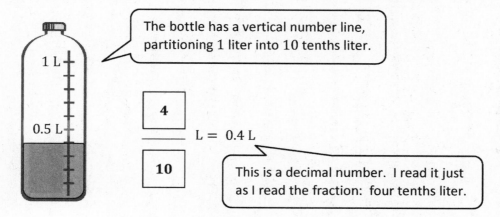

The bottle has a vertical number line, partitioning 1 liter into 10 tenths liter.

$$\frac{4}{10} \text{ L} = 0.4 \text{ L}$$

This is a decimal number. I read it just as I read the fraction: four tenths liter.

2. Write the weight of the pineapple on the scale in fraction form.

I can read the weight of the pineapple two ways: zero point nine kilograms or nine tenths kilogram.

$$\frac{9}{10} \text{ kg}$$

3. Fill in the blank to make the sentence true in both fraction form and decimal form.

$$\frac{3}{10} \text{ cm} + \underline{\frac{7}{10}} \text{ cm} = 1 \text{ cm}$$

$$0.3 \text{ cm} + \underline{\mathbf{0.7}} \text{ cm} = 1.0 \text{ cm}$$

$\frac{10}{10}$ cm is equal to 1 cm.

To find pairs of tenths that make 1.0 cm, I think of partners to 10, like 3 and 7, and 9 and 1.

EUREKA MATH®

Lesson 1: Use metric measurement to model the decomposition of one whole into tenths.

185

© 2018 Great Minds®. eureka-math.org

Name _____   Date _____

Shade the first 4 units of the tape diagram. Count by tenths to label the number line using a fraction and a decimal for each point. Circle the decimal that represents the shaded part.

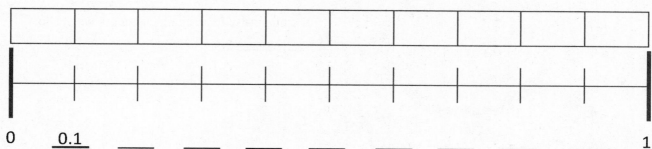

0   0.1   ___   ___   ___   ___   ___   ___   ___   ___   1
           $\frac{1}{10}$

2.   Write the total amount of water in fraction form and decimal form. Shade the last bottle to show the correct amount.

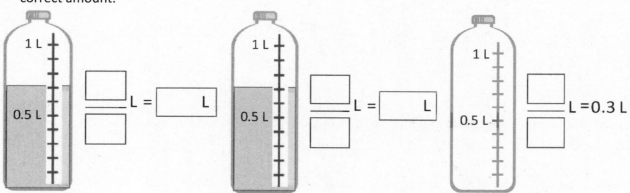

3.   Write the total weight of the food on each scale in fraction form or decimal form.

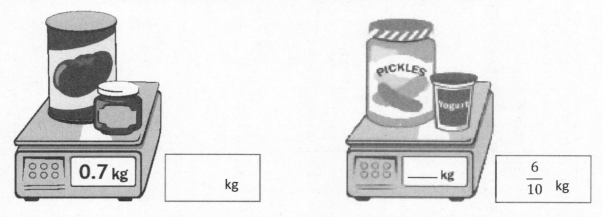

EUREKA MATH   **Lesson 1:**   Use metric measurement to model the decomposition of one whole   **187**
                                                          into tenths.

© 2018 Great Minds®. eureka-math.org

4. Write the length of the bug in centimeters. (The drawing is not to scale.)

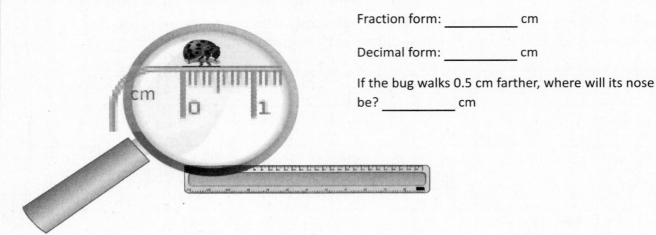

Fraction form: _____ cm

Decimal form: _____ cm

If the bug walks 0.5 cm farther, where will its nose be? _____ cm

5. Fill in the blank to make the sentence true in both fraction and decimal form.

    a. $\frac{4}{10}$ cm + _____ cm = 1 cm   0.4 cm + _____ cm = 1.0 cm

    b. $\frac{3}{10}$ cm + _____ cm = 1 cm   0.3 cm + _____ cm = 1.0 cm

    c. $\frac{8}{10}$ cm + _____ cm = 1 cm   0.8 cm + _____ cm = 1.0 cm

6. Match each amount expressed in unit form to its equivalent fraction and decimal.

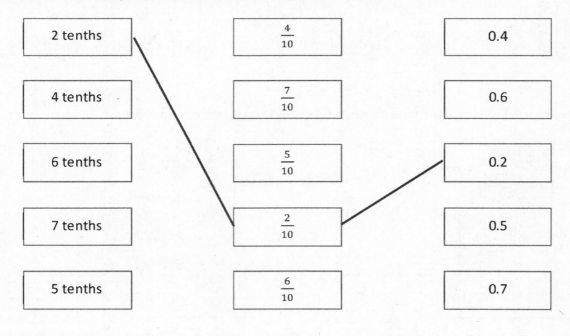

2 tenths		$\frac{4}{10}$		0.4
4 tenths		$\frac{7}{10}$		0.6
6 tenths		$\frac{5}{10}$		0.2
7 tenths		$\frac{2}{10}$		0.5
5 tenths		$\frac{6}{10}$		0.7

Lesson 1:   Use metric measurement to model the decomposition of one whole
                             into tenths.

EUREKA MATH®

1. For the length given below, draw a line segment to match. Express the measurement as an equivalent mixed number.

   2.7 cm

   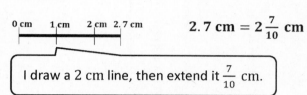

   2.7 cm = 2 $\frac{7}{10}$ cm

   I can express a decimal as a mixed number. The decimal and fractional part for this number have the unit *tenths*.

   I draw a 2 cm line, then extend it $\frac{7}{10}$ cm.

2. Write the following in decimal form. Then, model and rename the number.

   a. 1 one and 7 tenths = __**1.7**__

   Each rectangle represents 1. There are 10 tenths in 1.

   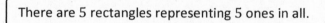

   I shade 17 tenths to show 1.7.

   $$1\frac{7}{10} = 1 + \frac{7}{10} = 1 + 0.7 = 1.7$$

   b. $\frac{22}{10}$ = __**2.2**__

   There are 5 rectangles representing 5 ones in all.

   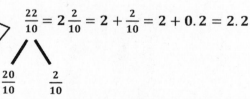

   I use a number bond to decompose the whole and the fraction. 20 tenths is equal to 2 ones.

   $$\frac{22}{10} = 2\frac{2}{10} = 2 + \frac{2}{10} = 2 + 0.2 = 2.2$$

    $\frac{20}{10}$ $\frac{2}{10}$

   How much more is needed to get to 5? __*2 ones 8 tenths*__

Name _____   Date _____

1. For each length given below, draw a line segment to match. Express each measurement as an equivalent mixed number.

   a. 2.6 cm

   b. 3.5 cm

   c. 1.7 cm

   d. 4.3 cm

   e. 2.2 cm

2. Write the following in decimal form. Then, model and rename the number as shown below.

   a. 2 ones and 4 tenths = _____

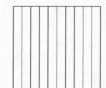   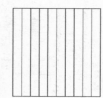 

$$2\frac{4}{10} = 2 + \frac{4}{10} = 2 + 0.4 = 2.4$$

EUREKA
MATH

Lesson 2:   Use metric measurement and area models to represent tenths as
              fractions greater than 1 and decimal numbers.

© 2018 Great Minds®. eureka-math.org

191

b. 3 ones and 8 tenths = _____

c. $4\frac{1}{10}$ = _____

   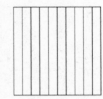

d. $1\frac{4}{10}$ = _____

 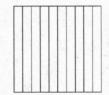   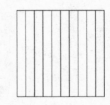

How much more is needed to get to 5? _____

e. $\frac{33}{10}$ = _____

    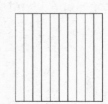

How much more is needed to get to 5? _____

Lesson 2: Use metric measurement and area models to represent tenths as fractions greater than 1 and decimal numbers.

EUREKA MATH®

1. Circle groups of tenths to make as many ones as possible.

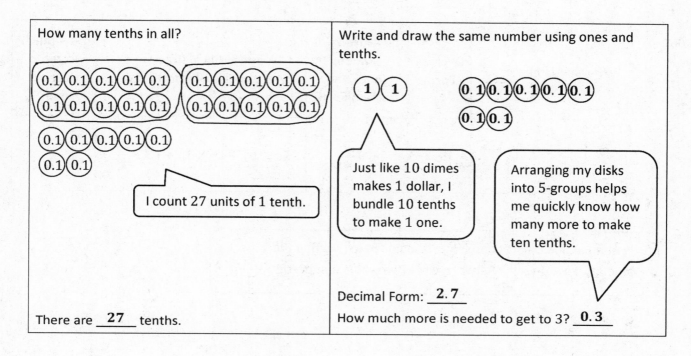

How many tenths in all?	Write and draw the same number using ones and tenths.

I count 27 units of 1 tenth.

There are __27__ tenths.

Just like 10 dimes makes 1 dollar, I bundle 10 tenths to make 1 one.

Arranging my disks into 5-groups helps me quickly know how many more to make ten tenths.

Decimal Form: __2.7__

How much more is needed to get to 3? __0.3__

2. Draw disks to represent 2 tens 3 ones 5 tenths using tens, ones, and tenths. Then, show the expanded form of the number in fraction form and decimal form.

$$(2 \times 10) + (3 \times 1) + \left(5 \times \frac{1}{10}\right) = 23\frac{5}{10}$$

I write a multiplication expression for the value of each digit in $23\frac{5}{10}$.

$$(2 \times 10) + (3 \times 1) + (5 \times 0.1) = 23.5$$

I can write in decimal form. Zero point one is another way to write 1 tenth.

EUREKA MATH®

Lesson 3: Represent mixed numbers with units of tens, ones, and tenths with place value disks, on the number line, and in expanded form.

© 2018 Great Minds®. eureka-math.org

193

3. Complete the chart.

Number Line	Decimal Form	Mixed Number (ones and fraction form)	Expanded Form (fraction or decimal form)	How much to get to the next one?
19 ●━━━━━━━━ 20	19.3	$19\frac{3}{10}$	$(1 \times 10) + (9 \times 1) + \left(3 \times \frac{1}{10}\right)$	$\frac{7}{10}$

The number line is partitioned into 10 equal parts. To find the endpoints, I ask myself, "Between what two whole numbers is $19\frac{3}{10}$?"

Lesson 3:   Represent mixed numbers with units of tens, ones, and tenths with place value disks, on the number line, and in expanded form.

EUREKA MATH

Name _____   Date _____

1. Circle groups of tenths to make as many ones as possible.

a. How many tenths in all?	Write and draw the same number using ones and tenths.
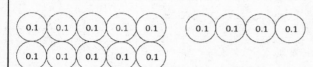 There are _____ tenths.	Decimal Form: _____ How much more is needed to get to 2? _____

b. How many tenths in all?	Write and draw the same number using ones and tenths.
There are _____ tenths.	Decimal Form: _____ How much more is needed to get to 3? _____

2. Draw disks to represent each number using tens, ones, and tenths. Then, show the expanded form of the number in fraction form and decimal form as shown. The first one has been completed for you.

a. 3 tens 4 ones 3 tenths	b. 5 tens 3 ones 7 tenths
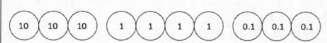 Fraction Expanded Form $(3 \times 10) + (4 \times 1) + (3 \times \frac{1}{10}) = 34\frac{3}{10}$ Decimal Expanded Form $(3 \times 10) + (4 \times 1) + (3 \times 0.1) = 34.3$	

EUREKA MATH

Lesson 3:   Represent mixed numbers with units of tens, ones, and tenths with place value disks, on the number line, and in expanded form.   195

© 2018 Great Minds®. eureka-math.org

c. 3 tens 2 ones 3 tenths	d. 8 tens 4 ones 8 tenths

3. Complete the chart.

Point	Number Line	Decimal Form	Mixed Number (ones and fraction form)	Expanded Form (fraction or decimal form)	How much to get to the next one?
a.			$4\frac{6}{10}$		
b.	24 25				0.5
c.				$(6\times10)+(3\times1)+(6\times\frac{1}{10})$	
d.			$71\frac{3}{10}$		
e.				$(9\times10)+(9\times0.1)$	

Lesson 3:   Represent mixed numbers with units of tens, ones, and tenths with
                        place value disks, on the number line, and in expanded form.

EUREKA
MATH

1 meter equals 100 centimeters. When a meter is decomposed into 10 equal parts, 1 part equals $\frac{1}{10}$ meter or 10 centimeters.

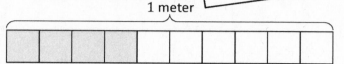

1 meter

1.

a. What is the length of the shaded part of the meter stick in centimeters?

40 *centimeters*

b. What fraction of a meter is 4 centimeters?

$\frac{4}{100}$ *meter*

Each tenth of a meter would need to be decomposed into 10 equal parts to show all 100 centimeters in 1 meter. To represent 4 centimeters, I would shade 4 of the 100 parts.

c. What fraction of a meter is 40 centimeters?

$\frac{4}{10}$ *meter or* $\frac{40}{100}$ *meter*

1 meter

2. Fill in the blank.

$$\frac{3}{10}\text{m} = \frac{30}{100}\text{m}$$

1 out of 100 centimeters is 1 hundredth centimeter.

3. On the meter stick, shade in the amount shown. Then, write the equivalent decimal.

$\frac{51}{100}$m = **0.51 m**

1 meter

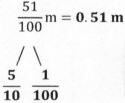

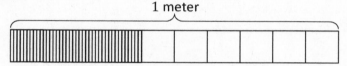

I shade 5 tenths of a meter. After partitioning the next tenth meter into 10 equal parts, I shade 1 hundredth meter more.

4. Draw a number bond, pulling out the tenths from the hundredths. Write the total as the equivalent decimal.

$\frac{87}{100}$

0.87

8 tenths is the same as 80 hundredths.

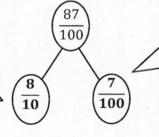

I can decompose a fraction like I decompose a whole number. I break 87 hundredths into 80 hundredths and 7 hundredths.

EUREKA MATH   Lesson 4:   Use meters to model the decomposition of one whole into hundredths.   **197**
                                  Represent and count hundredths.

© 2018 Great Minds®. eureka-math.org

Name _____   Date _____

1. a. What is the length of the shaded part of the meter stick in centimeters?

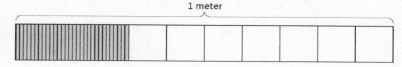

   b. What fraction of a meter is 3 centimeters?

   c. In fraction form, express the length of the shaded portion of the meter stick.

   d. In decimal form, express the length of the shaded portion of the meter stick.

   e. What fraction of a meter is 30 centimeters?

2. Fill in the blanks.

   a. 5 tenths = ____ hundredths

   b. $\frac{5}{10}$ m = $\frac{}{100}$ m

   c. $\frac{4}{10}$ m = $\frac{40}{}$ m

3. Use the model to add the shaded parts as shown. Write a number bond with the total written in decimal form and the parts written as fractions. The first one has been done for you.

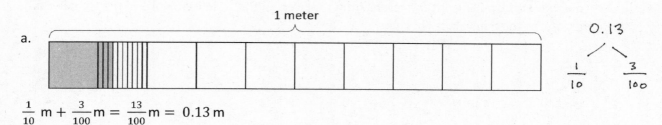

   a.

   $\frac{1}{10}$ m + $\frac{3}{100}$ m = $\frac{13}{100}$ m = 0.13 m

Lesson 4:   Use meters to model the decomposition of one whole into hundredths. Represent and count hundredths.

© 2018 Great Minds®. eureka-math.org

199

b.

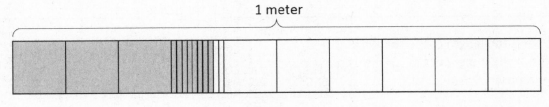

c.

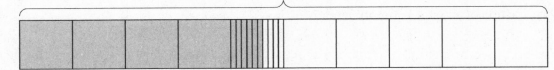

4. On each meter stick, shade in the amount shown. Then, write the equivalent decimal.

a. $\frac{9}{10}$ m

1 meter

b. $\frac{15}{100}$ m

1 meter

c. $\frac{41}{100}$ m

1 meter

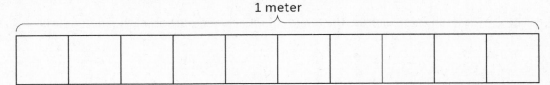

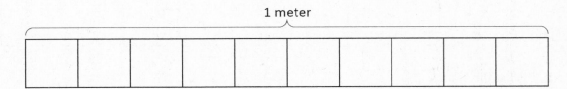

5. Draw a number bond, pulling out the tenths from the hundredths, as in Problem 3 of the Homework.
    Write the total as the equivalent decimal.

a. $\frac{23}{100}$ m

b. $\frac{38}{100}$ m

c. $\frac{82}{100}$

d. $\frac{76}{100}$

Lesson 4:   Use meters to model the decomposition of one whole into hundredths.
                                 Represent and count hundredths.

EUREKA MATH

1. Find the equivalent fraction using multiplication or division. Shade the area models to show the equivalency. Record it as a decimal.

   a. $\dfrac{1 \times 10}{10 \times 10} = \dfrac{10}{100}$

   > I multiply the number of tenths by 10 to get the number of hundredths.

   b. $\dfrac{70 \div 10}{100 \div 10} = \dfrac{7}{10}$

   > I divide the number of hundredths by 10 to get the number of tenths.

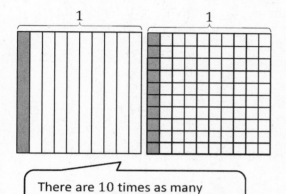

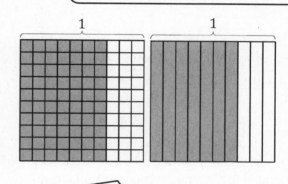

   > There are 10 times as many hundredths as there are tenths.

   > $\dfrac{7}{10}$ and $\dfrac{70}{100}$ are equivalent fractions.

2. Complete the number sentence. Shade the equivalent amount on the area model, drawing horizontal lines to make hundredths.

   a. 25 hundredths = __**2**__ tenths + __**5**__ hundredths

   b. Decimal Form: __**0.25**__

   c. Fraction Form: $\dfrac{25}{100}$

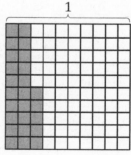

EUREKA MATH

Lesson 5:   Model the equivalence of tenths and hundredths using the area model and place value disks.

201

© 2018 Great Minds®. eureka-math.org

3. Circle hundredths to compose as many tenths as you can. Complete the number sentence. Represent the composition with a number bond.

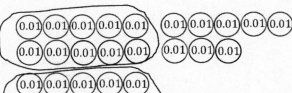

___28___ hundredths = ___2___ tenths + ___8___ hundredths

I compose 10 hundredths to make 1 tenth because $\frac{1}{10} = \frac{10}{100}$.

4. Use both tenths and hundredths place value disks to represent each number. Write the equivalent number in decimal, fraction, and unit form.

a. $\frac{54}{100} = 0.54$

(0.1)(0.1)(0.1)(0.1)(0.1)   (0.01)(0.01)(0.01)(0.01)

___54___ hundredths

b. $\frac{60}{100} = \mathbf{0.60}$

(0.1)(0.1)(0.1)(0.1)(0.1)
(0.1)

Since I know that $\frac{6}{10} = \frac{60}{100}$, it is more efficient to show 6 tenths than 60 hundredths.

60 hundredths

Lesson 5: Model the equivalence of tenths and hundredths using the area model and place value disks.

EUREKA MATH

Name _____   Date _____

1. Find the equivalent fraction using multiplication or division. Shade the area models to show the equivalency. Record it as a decimal.

a. $\dfrac{4 \times}{10 \times} = \dfrac{}{100}$

b. $\dfrac{60 \div}{100 \div} = \dfrac{}{10}$

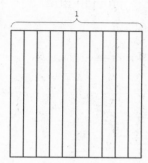
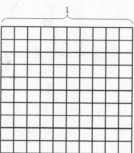
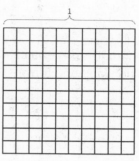

2. Complete the number sentences. Shade the equivalent amount on the area model, drawing horizontal lines to make hundredths.

a. 36 hundredths = _____ tenths + _____ hundredths

Decimal form: _____

Fraction form: _____

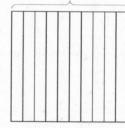

b. 82 hundredths = _____ tenths + _____ hundredths

Decimal form: _____

Fraction form: _____

3. Circle hundredths to compose as many tenths as you can. Complete the number sentences. Represent each with a number bond as shown.

a.

0.01 0.01 0.01 0.01 0.01   0.01 0.01 0.01 0.01

0.01 0.01 0.01 0.01 0.01

$$0.14$$
$$\swarrow \qquad \searrow$$
$$\frac{1}{10} \qquad \frac{4}{100}$$

_____ hundredths = _____ tenth + _____ hundredths

EUREKA
MATH

Lesson 5:   Model the equivalence of tenths and hundredths using the area model and place value disks.

203

© 2018 Great Minds®. eureka-math.org

b.

(0.01) (0.01) (0.01) (0.01) (0.01)   (0.01) (0.01) (0.01) (0.01)

(0.01) (0.01) (0.01) (0.01) (0.01)

(0.01) (0.01) (0.01) (0.01) (0.01)

(0.01) (0.01) (0.01) (0.01) (0.01)   _____ hundredths = _____ tenths + _____ hundredths

4. Use both tenths and hundredths place value disks to represent each number. Write the equivalent number in decimal, fraction, and unit form.

a. $\frac{4}{100}$ = 0. _____ _____ hundredths	b. $\frac{13}{100}$ = 0. _____ _____ tenth _____ hundredths
c. ⎯⎯ = 0.41 _____ hundredths	d. ⎯⎯ = 0.90 _____ tenths
e. ⎯⎯ = 0. _____ 6 tenths 3 hundredths	f. ⎯⎯ = 0. _____ 90 hundredths

Lesson 5:   Model the equivalence of tenths and hundredths using the area model and place value disks.

EUREKA MATH®

1. Shade the area models to represent the number, drawing horizontal lines to make hundredths as needed. Locate the corresponding point on the number line. Label with a point, and record the mixed number as a decimal.

$3\frac{42}{100} = $ **3.42**

There are 3 ones in $3\frac{42}{100}$. I shade 3 area models completely.

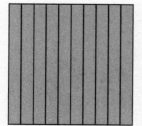  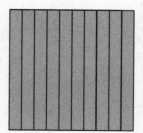 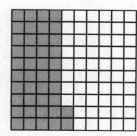

I shade 42 hundredths after drawing horizontal lines to decompose tenths into hundredths.

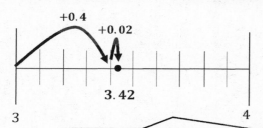

+0.4
+0.02
3.42
3   4

To find 3.42 on the number line, I begin with the largest unit. I start at 3 ones. I slide 4 tenths. Then, I estimate where 2 hundredths would be.

2. Write the equivalent fraction and decimal for the following number.

9 ones 7 hundredths

$9\frac{7}{100}$ **9.07**

There are no tenths in this number! I show that with a zero as a placeholder.

To write a decimal number, I place a decimal point between the ones and the fraction.

EUREKA MATH®

Lesson 6: Use the area model and number line to represent mixed numbers with units of ones, tenths, and hundredths in fraction and decimal forms.

205

© 2018 Great Minds®. eureka-math.org

Name _____ Date _____

1. Shade the area models to represent the number, drawing horizontal lines to make hundredths as needed. Locate the corresponding point on the number line. Label with a point, and record the mixed number as a decimal.

   a. $2\frac{35}{100}$ = ___.___

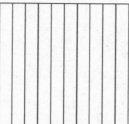 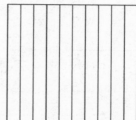 

   b. $3\frac{17}{100}$ = ___.___

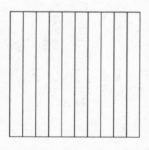 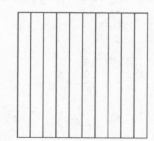 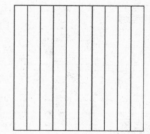 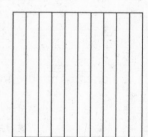

2. Estimate to locate the points on the number lines.

   a. $5\frac{90}{100}$

   b. $3\frac{25}{100}$

Lesson 6: Use the area model and number line to represent mixed numbers with units of ones, tenths, and hundredths in fraction and decimal forms.

207

EUREKA MATH

© 2018 Great Minds®. eureka-math.org

3. Write the equivalent fraction and decimal for each of the following numbers.

a. 2 ones 2 hundredths	b. 2 ones 16 hundredths
c. 3 ones 7 hundredths	d. 1 one 18 hundredths
e. 9 ones 62 hundredths	f. 6 ones 20 hundredths

4. Draw lines from dot to dot to match the decimal form to both the unit form and fraction form. All unit forms and fractions have at least one match, and some have more than one match.

4 ones 18 hundredths ●

4 ones 8 hundredths ●

4 ones 8 tenths ●

4 tens 8 ones ●

● 4.80 ●

● 4.8 ●

● 4.18 ●

● 4.08 ●

● 48 ●

● $4\frac{18}{100}$

● 48

● $4\frac{8}{100}$

● $4\frac{80}{100}$

Lesson 6:   Use the area model and number line to represent mixed numbers with units of ones, tenths, and hundredths in fraction and decimal forms.

EUREKA MATH®

1. Write a decimal number sentence to identify the total value of the place value disks.

(10)(10)   (1)   (0.1)(0.1)(0.1)(0.1)(0.1)   (0.01)(0.01)(0.01)(0.01)

2 tens   1 one   5 tenths   4 hundredths

__20__ + __1__ + __0.5__ + __0.04__ = __21.54__

> I write the expanded form.

2. Use the place value chart to answer the following questions. Express the value of the digit in unit form.

hundreds	tens	ones	.	tenths	hundredths
3	5	1	.	8	2

a. The digit __3__ is in the hundreds place. It has a value of __3 hundreds__

> I write the value of 300 in unit form.

b. The digit __5__ is in the tens place. It has a value of __5 tens__.

3. Write the decimal as an equivalent fraction. Then, write the number in expanded form, using both decimal and fraction notation.

Decimal and Fraction Form	Expanded Form	
	Fraction Notation	Decimal Notation
$27.03 = 27\frac{3}{100}$	$(2 \times 10) + (7 \times 1) + \left(3 \times \frac{1}{100}\right)$ $20 + 7 + \frac{3}{100}$	$(2 \times 10) + (7 \times 1) + (3 \times 0.01)$ $20 + 7 + 0.03$
$400.80 = 400\frac{80}{100}$	$(4 \times 100) + \left(8 \times \frac{1}{10}\right)$ $400 + \frac{8}{10}$	$(4 \times 100) + (8 \times 0.1)$ $400 + 0.8$

> This number has many zeros! There are values in the hundreds and tenths place that I show as addends in the expressions.

> Expanded form can be written two ways. Using parentheses, I show how the value of each digit is a multiple of a base-ten unit (e.g., 4×100). Or, I show the value of each digit (e.g., 400).

Name _____   Date _____

1. Write a decimal number sentence to identify the total value of the place value disks.

    a.

    3 tens   4 tenths   2 hundredths

    _____ + _____ + _____ = _____

    b.

    4 hundreds   3 hundredths

    _____ + _____ = _____

2. Use the place value chart to answer the following questions. Express the value of the digit in unit form.

hundreds	tens	ones	.	tenths	hundredths
8	2	7		6	4

a. The digit _____ is in the hundreds place. It has a value of _____.

b. The digit _____ is in the tens place. It has a value of _____.

c. The digit _____ is in the tenths place. It has a value of _____.

d. The digit _____ is in the hundredths place. It has a value of _____.

hundreds	tens	ones	.	tenths	hundredths
3	4	5		1	9

e. The digit _____ is in the hundreds place. It has a value of _____.

f. The digit _____ is in the tens place. It has a value of _____.

g. The digit _____ is in the tenths place. It has a value of _____.

h. The digit _____ is in the hundredths place. It has a value of _____.

EUREKA MATH®

Lesson 7:   Model mixed numbers with units of hundreds, tens, ones, tenths, and
             hundredths in expanded form and on the place value chart.

© 2018 Great Minds®. eureka-math.org

211

3. Write each decimal as an equivalent fraction. Then, write each number in expanded form, using both decimal and fraction notation. The first one has been done for you.

Decimal and Fraction Form	Expanded Form	
	Fraction Notation	Decimal Notation
$14.23 = 14\frac{23}{100}$	$(1 \times 10) + (4 \times 1) + (2 \times \frac{1}{10}) + (3 \times \frac{1}{100})$ $10 \quad + \quad 4 \quad + \quad \frac{2}{10} \quad + \quad \frac{3}{100}$	$(1 \times 10) + (4 \times 1) + (2 \times 0.1) + (3 \times 0.01)$ $10 \quad + \quad 4 \quad + \quad 0.2 \quad + \quad 0.03$
25.3 = _____		
39.07 = _____		
40.6 = _____		
208.90 = _____		
510.07 = _____		
900.09 = _____		

Lesson 7: Model mixed numbers with units of hundreds, tens, ones, tenths, and hundredths in expanded form and on the place value chart.

EUREKA MATH

1. Use the area model to represent $\frac{140}{100}$. Complete the number sentence.

$\frac{140}{100}$ = __14__ tenths = __1__ one __4__ tenths = __1.4__

 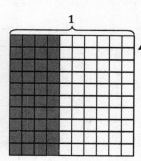

> I can draw horizontal lines to show hundredths. 1 one equals 10 tenths or 100 hundredths. 4 tenths equals 40 hundredths.

> I shade 14 tenths. My model shows that 14 tenths is the same as 1 one and 4 tenths.

2. Draw place value disks to represent the following decomposition:

2 tenths 3 hundredths = __23__ hundredths

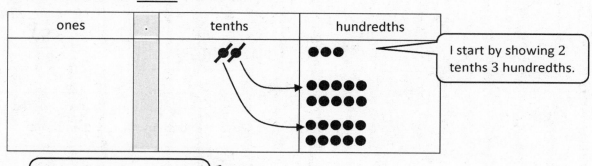

> I start by showing 2 tenths 3 hundredths.

> I decompose 2 tenths as 20 hundredths.

EUREKA MATH

Lesson 8: Use understanding of fraction equivalence to investigate decimal numbers on the place value chart expressed in different units.

213

© 2018 Great Minds®. eureka-math.org

3. Decompose the units to represent each number as tenths.

   a. 1.3 = __13__ tenths

   b. 18.3 = __183__ tenths

4. Decompose the units to represent each number as hundredths.

   a. 1.3 = __130__ hundredths

   b. 18.3 = __1,830__ hundredths

   > I notice a pattern! There are 10 times as many hundredths as tenths.

5. Complete the chart.

Decimal	Mixed Number	Tenths	Hundredths
8.2	$8\frac{2}{10}$	82 *tenths* $\frac{82}{10}$	820 *hundredths* $\frac{820}{100}$

> I write tenths and hundredths in both fraction and unit form.

Lesson 8: Use understanding of fraction equivalence to investigate decimal numbers on the place value chart expressed in different units.

© 2018 Great Minds®. eureka-math.org

EUREKA MATH

Name _____   Date _____

1. Use the area model to represent $\frac{220}{100}$. Complete the number sentence.

   a. $\frac{220}{100}$ = _____ tenths = _____ ones _____ tenths = ___._____

    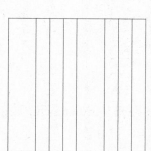 

   b. In the space below, explain how you determined your answer to part (a).

2. Draw place value disks to represent the following decompositions:

   3 ones = _____ tenths

ones	.	tenths	hundredths

   3 tenths = _____ hundredths

ones	.	tenths	hundredths

   2 ones 3 tenths = _____ tenths

ones	.	tenths	hundredths

   3 tenths 3 hundredths = _____ hundredths

ones	.	tenths	hundredths

EUREKA MATH®

Lesson 8:   Use understanding of fraction equivalence to investigate decimal
numbers on the place value chart expressed in different units.

215

© 2018 Great Minds®. eureka-math.org

3. Decompose the units to represent each number as tenths.

a. 1 = _____ tenths

b. 2 = _____ tenths

c. 1.3 = _____ tenths

d. 2.6 = _____ tenths

e. 10.3 = _____ tenths

f. 20.6 = _____ tenths

4. Decompose the units to represent each number as hundredths.

a. 1 = _____ hundredths

b. 2 = _____ hundredths

c. 1.3 = _____ hundredths

d. 2.6 = _____ hundredths

e. 10.3 = _____ hundredths

f. 20.6 = _____ hundredths

5. Complete the chart. The first one has been done for you.

Decimal	Mixed Number	Tenths	Hundredths
4.1	$4\frac{1}{10}$	41 tenths $\frac{41}{10}$	410 hundredths $\frac{410}{100}$
5.3			
9.7			
10.9			
68.5			

Lesson 8: Use understanding of fraction equivalence to investigate decimal numbers on the place value chart expressed in different units.

© 2018 Great Minds®. eureka-math.org

EUREKA MATH®

1. Express the lengths of the shaded parts in decimal form. Write a sentence that compares the two lengths. Use the expression *shorter than* or *longer than* in your sentence.

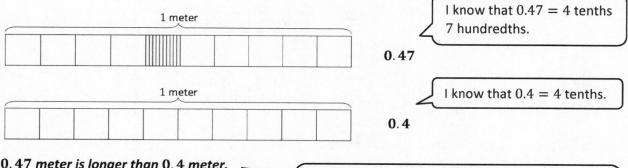

1 meter

0.47

> I know that 0.47 = 4 tenths 7 hundredths.

1 meter

0.4

> I know that 0.4 = 4 tenths.

0.47 *meter is longer than* 0.4 *meter.*

> Both numbers have 4 tenths. 0.47 meter is longer because it has an additional 7 hundredths. I can see that by looking at the tape diagrams.

2. Examine the mass of each item as shown below on the 1-kilogram scales. Put an X over the items that are lighter than the bananas.

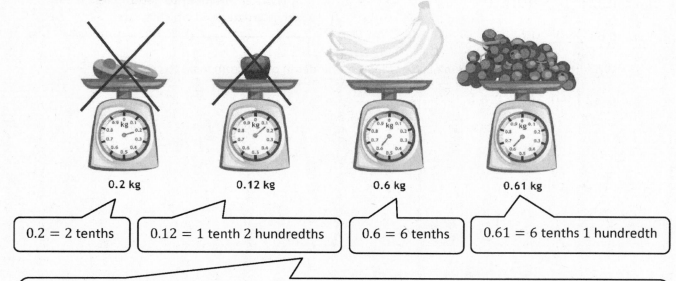

| 0.2 kg | 0.12 kg | 0.6 kg | 0.61 kg |

> 0.2 = 2 tenths

> 0.12 = 1 tenth 2 hundredths

> 0.6 = 6 tenths

> 0.61 = 6 tenths 1 hundredth

> I compare by looking at the largest place value unit in the mass of each item. The largest unit in each item is tenths. The avocado and the apple have fewer tenths than the bananas. The grapes have the same number of tenths, but they also have 1 more hundredth. The grapes are heavier than the bananas.

EUREKA MATH

Lesson 9: Use the place value chart and metric measurement to compare decimals and answer comparison questions.

217

© 2018 Great Minds®. eureka-math.org

3. Record the volume of water in each graduated cylinder on the place value chart below.

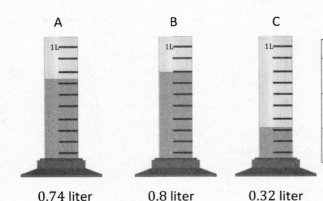

A   B   C

0.74 liter   0.8 liter   0.32 liter

Volume of Water (in liters)

Cylinder	ones	.	tenths	hundredths
A	0	.	7	4
B	0	.	8	0
C	0	.	3	2

Compare the values using >, <, or =.

a.   0.74 L __>__ 0.32 L

b.   0.32 L __<__ 0.8 L

c.   0.8 L __>__ 0.74 L

> I look at the pictures and the completed table to help me compare the values. Tenths are the largest unit in each number, so I can compare the number of tenths in each number to determine which is greater and which is less.

d.   Write the volume of water in each graduated cylinder in order from least to greatest.

0.32 L,   0.74 L,   0.8 L

Lesson 9: Use the place value chart and metric measurement to compare decimals and answer comparison questions.

© 2018 Great Minds®. eureka-math.org

EUREKA
MATH

Name _____ Date _____

1. Express the lengths of the shaded parts in decimal form. Write a sentence that compares the two lengths. Use the expression *shorter than* or *longer than* in your sentence.

a.

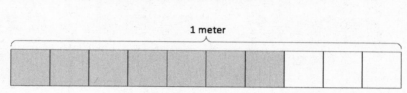

b.

c. List all four lengths from least to greatest.

Lesson 9: Use the place value chart and metric measurement to compare decimals and answer comparison questions.

© 2018 Great Minds®. eureka-math.org

EUREKA MATH

219

2. a. Examine the mass of each item as shown below on the 1-kilogram scales. Put an X over the items
        that are heavier than the volleyball

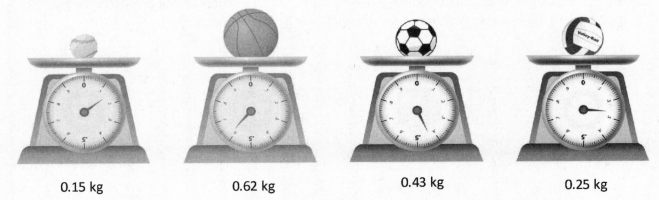

| 0.15 kg | 0.62 kg | 0.43 kg | 0.25 kg |

b. Express the mass of each item on the place value chart.

Mass of Sport Balls (kilograms)

Sport Balls	ones	.	tenths	hundredths
baseball				
volleyball				
basketball				
soccer ball				

c. Complete the statements below using the words *heavier than* or *lighter than* in your statements.

The soccer ball is _____ the baseball.

The volleyball is _____ the basketball.

Lesson 9: Use the place value chart and metric measurement to compare
decimals and answer comparison questions.

EUREKA
MATH

3. Record the volume of water in each graduated cylinder on the place value chart below.

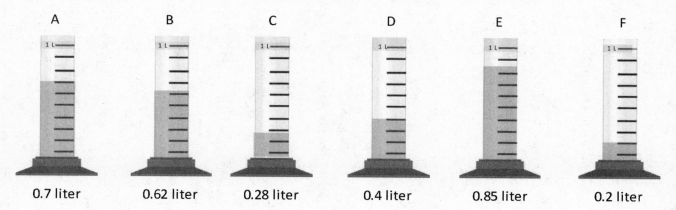

A	B	C	D	E	F
0.7 liter	0.62 liter	0.28 liter	0.4 liter	0.85 liter	0.2 liter

Volume of Water (liters)

Cylinder	ones	.	tenths	hundredths
A				
B				
C				
D				
E				
F				

Compare the values using >, <, or =.

a.   0.4 L _____ 0.2 L

b.   0.62 L _____ 0.7 L

c.   0.2 L _____ 0.28 L

d.   Write the volume of water in each graduated cylinder in order from least to greatest.

Lesson 9:   Use the place value chart and metric measurement to compare decimals and answer comparison questions.

221

1. Shade the area models below, decomposing tenths as needed, to represent the pair of decimal numbers. Fill in the blank with <, >, or = to compare the decimal numbers.

   0.4 __>__ 0.37

   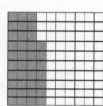

   At first, I thought, "37 is greater than 4." But then I remembered the units of these numbers must be the same in order to compare. 4 tenths is equal to 40 hundredths, and 40 hundredths is greater than 37 hundredths.

2. Locate and label the points for each of the decimal numbers on the number line. Fill in the blank with <, >, or = to compare the decimal numbers.

   11.02 __<__ 11.21

   Each tick mark represents 1 hundredth. 11.0 equals 11 and 0 hundredths. 11.02 equals 11 and 2 hundredths. 11.21 equals 11 and 21 hundredths. I use this information to help me to locate and label the points.

3. Use the symbols <, >, or = to compare.

   1.7 __>__ 1.17

   I know that 1.7 is greater than 1.17 because 1.7 = 1.70 and 1.70 > 1.17.

4. Use the symbols <, >, or = to compare. Use a picture as needed to solve.

   47 tenths __>__ 4.6

   I rename 47 tenths as 4 and 7 tenths. 4.7 > 4.6

Name _____   Date _____

1. Shade the parts of the area models below, decomposing tenths as needed, to represent the pairs of decimal numbers. Fill in the blank with <, >, or = to compare the decimal numbers.

    a. 0.19 _____ 0.3   b. 0.6 _____ 0.06

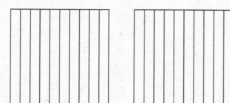   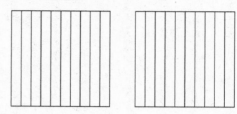

    c. 1.8 _____ 1.53   d. 0.38 _____ 0.7

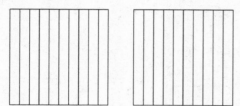   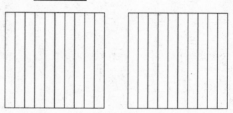

2. Locate and label the points for each of the decimal numbers on the number line.
    Fill in the blank with <, >, or = to compare the decimal numbers.

    a. 7.2 _____ 7.02

    b. 18.19 _____ 18.3

EUREKA MATH®

© 2018 Great Minds®. eureka-math.org

3. Use the symbols <, >, or = to compare.

   a. 2.68 _____ 2.54

   b. 6.37 _____ 6.73

   c. 9.28 _____ 7.28

   d. 3.02 _____ 3.2

   e. 13.1 _____ 13.10

   f. 5.8 _____ 5.92

4. Use the symbols <, >, or = to compare. Use pictures as needed to solve.

   a. 57 tenths _____ 5.7

   b. 6.2 _____ 6 ones and 2 hundredths

   c. 33 tenths _____ 33 hundredths

   d. 8.39 _____ $8\frac{39}{10}$

   e. $\frac{236}{100}$ _____ 2.36

   f. 3 tenths _____ 22 hundredths

Lesson 10:   Use area models and the number line to compare decimal numbers, and record comparisons using <, >, and =.

© 2018 Great Minds®. eureka-math.org

EUREKA MATH

1. Plot the following points on the number line.

1.56, $1\frac{6}{10}$, $\frac{163}{100}$, $\frac{17}{10}$, 1.62, 1 one and 75 hundredths

$1\frac{56}{100}$ $1\frac{60}{100}$ $1\frac{63}{100}$ $1\frac{70}{100}$ $1\frac{62}{100}$ $1\frac{75}{100}$

> I rename all of the numbers to fractions with like units—hundredths. I know that each tick mark represents 1 hundredth.

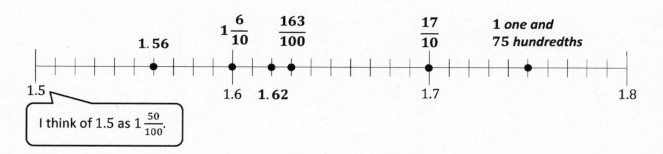

> I think of 1.5 as $1\frac{50}{100}$.

2. Arrange the following numbers in order from greatest to least using decimal form. Use the > symbol between each number.

7 ones and 23 hundredths, $\frac{725}{100}$, 7.4, $7\frac{52}{100}$, $8\frac{2}{10}$, $7\frac{4}{100}$

$8.2 > 7.52 > 7.4 > 7.25 > 7.23 > 7.04$

> I rename all of the numbers to decimal form. To help me order the numbers, I think of $8\frac{2}{10}$ as 8.20 and 7.4 as 7.40.

3. In a frog-jumping contest, Mary's frog jumped 1.04 meters. Kelly's frog jumped 1.4 meters, and Katrina's frog jumped 1.14 meters. Whose frog jumped the farthest distance? Whose frog jumped the shortest distance?

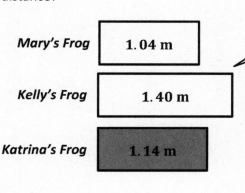

> I rename 1.4 to 1.40 to be able to compare hundredths.

Kelly's frog jumped the farthest distance. Mary's frog jumped the shortest distance. I know because they all jumped at least 1 meter, but Kelly's frog jumped an additional 40 hundredths meter, and Mary's frog only jumped an additional 4 hundredths meter.

Name _____   Date _____

1. Plot the following points on the number line using decimal form.

   a. 0.6, $\frac{5}{10}$, 0.76, $\frac{79}{100}$, 0.53, $\frac{67}{100}$

   0.5   0.6   0.7   0.8

   b. 8 ones and 15 hundredths, $\frac{832}{100}$, $8\frac{27}{100}$, $\frac{82}{10}$, 8.1

   8.1   8.2   8.3   8.4

   c. $13\frac{12}{100}$, $\frac{130}{10}$, 13 ones and 3 tenths, 13.21, $13\frac{3}{100}$

   13.0   13.1   13.2   13.3

EUREKA MATH

Lesson 11:   Compare and order mixed numbers in various forms.

229

© 2018 Great Minds®. eureka-math.org

2. Arrange the following numbers in order from greatest to least using decimal form. Use the > symbol between each number.

a. 4.03, 4 ones and 33 hundredths, $\frac{34}{100}$, $4\frac{43}{100}$, $\frac{430}{100}$, 4.31

_____

b. $17\frac{5}{10}$, 17.55, $\frac{157}{10}$, 17 ones and 5 hundredths, 15.71, $15\frac{75}{100}$

_____

c. 8 ones and 19 hundredths, $9\frac{8}{10}$, 81, $\frac{809}{100}$, 8.9, $8\frac{1}{10}$

_____

3. In a paper airplane contest, Matt's airplane flew 9.14 meters. Jenna's airplane flew $9\frac{4}{10}$ meters. Ben's airplane flew $\frac{904}{100}$ meters. Leah's airplane flew 9.1 meters. Whose airplane flew the farthest?

4. Becky drank $1\frac{41}{100}$ liters of water on Monday, 1.14 liters on Tuesday, 1.04 liters on Wednesday, $\frac{11}{10}$ liters on Thursday, and $1\frac{40}{100}$ liters on Friday. Which day did Becky drink the most? Which day did Becky drink the least?

EUREKA MATH

1. Complete the number sentence by expressing each part using hundredths. Model using the place value chart.

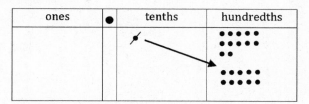

ones	•	tenths	hundredths

1 tenth + 12 hundredths = __22__ hundredths

10 hundredths + 12 hundredths = 22 hundredths

> To make like units, I change 1 tenth to 10 hundredths.
> 10 hundredths + 12 hundredths = 22 hundredths.

2. Solve by converting all addends to hundredths before solving.

   a.   6 tenths + 21 hundredths = __60__ hundredths + __21__ hundredths = __81__ hundredths

   > This is just like Problem 1. Instead of drawing place value disks, I change the tenths to hundredths in my mind. Each tenth equals 10 hundredths.

   b.   27 hundredths + 3 tenths = __27__ hundredths + __30__ hundredths = __57__ hundredths

   > I can't add because the units are not alike. I can't add 1 cat plus 2 dogs; I have to rename with like units. I can add 1 animal plus 2 animals.

3. Solve. Write your answer as a decimal.

   a. $\frac{3}{10} + \frac{21}{100}$

   $\frac{30}{100} + \frac{21}{100} = \frac{51}{100} = 0.51$

   b. $\frac{14}{100} + \frac{7}{10}$

   $\frac{14}{100} + \frac{70}{100} = \frac{84}{100} = 0.84$

   > To solve, I make like units of hundredths. I add, and then I change the answer from fraction form to decimal form.

EUREKA MATH

Lesson 12:   Apply understanding of fraction equivalence to add tenths and hundredths.

© 2018 Great Minds®. eureka-math.org

231

Name _____ Date _____

1. Complete the number sentence by expressing each part using hundredths. Model using the place value chart, as shown in part (a).

ones		tenths	hundredths

a. 1 tenth + 8 hundredths = _____ hundredths

ones		tenths	hundredths

b. 2 tenths + 3 hundredths = _____ hundredths

ones		tenths	hundredths

c. 1 tenth + 14 hundredths = _____ hundredths

2. Solve by converting all addends to hundredths before solving.

a. 1 tenth + 2 hundredths = _____ hundredths + 2 hundredths = _____ hundredths

b. 4 tenths + 11 hundredths = _____ hundredths + _____ hundredths = _____ hundredths

c. 8 tenths + 25 hundredths = _____ hundredths + _____ hundredths = _____ hundredths

d. 43 hundredths + 6 tenths = _____ hundredths + _____ hundredths = _____ hundredths

EUREKA MATH®

Lesson 12: Apply understanding of fraction equivalence to add tenths and hundredths.

© 2018 Great Minds®. eureka-math.org

233

3. Find the sum. Convert tenths to hundredths as needed. Write your answer as a decimal.

   a. $\frac{3}{10} + \frac{7}{100}$

   b. $\frac{16}{100} + \frac{5}{10}$

   c. $\frac{5}{10} + \frac{40}{100}$

   d. $\frac{20}{100} + \frac{8}{10}$

4. Solve. Write your answer as a decimal.

   a. $\frac{5}{10} + \frac{53}{100}$

   b. $\frac{27}{100} + \frac{8}{10}$

   c. $\frac{4}{10} + \frac{78}{100}$

   d. $\frac{98}{100} + \frac{7}{10}$

5. Cameron measured $\frac{65}{100}$ inch of rainwater on the first day of April. On the second day of April, he measured $\frac{83}{100}$ inch of rainwater. How many total inches of rainwater did Cameron measure on the first two days of April?

Lesson 12:   Apply understanding of fraction equivalence to add tenths and hundredths.

© 2018 Great Minds®. eureka-math.org

EUREKA MATH

Lesson Notes

In Grade 4, students add decimals by first writing the addends in fraction form and then adding the fractions to find the total. This strengthens student understanding of the fraction and decimal relationship, increases their ability to think flexibly, and prepares them for greater success with fractions and decimals in Grade 5.

1. Solve. Convert tenths to hundredths before finding the sum. Rewrite the complete number sentence in decimal form.

   a. $2\frac{31}{100} + \frac{4}{10}$

   > I convert 4 tenths to 40 hundredths. I add like units.

   $$2\frac{31}{100} + \frac{4}{10} = 2\frac{31}{100} + \frac{40}{100} = 2\frac{71}{100}$$

   > Decimal form is another way to express the numbers.

   $$2.31 + 0.40 = 2.71$$

   b. $4\frac{42}{100} + 2\frac{7}{10}$

   > I add ones to ones and hundredths to hundredths.

   $$4\frac{42}{100} + 2\frac{7}{10} = 4\frac{42}{100} + 2\frac{70}{100} = 6\frac{112}{100} = 7\frac{12}{100}$$

   $$1 \quad \frac{12}{100}$$

   > I use a number bond to show $\frac{112}{100} = 1 + \frac{12}{100}$ since $\frac{100}{100} = 1$.

   $$4.42 + 2.70 = 7.12$$

2. Solve by rewriting the expression in fraction form. After solving, rewrite the complete number sentence in decimal form.

   $4.4 + 1.74$

   > To add decimal numbers, I solve by relating this problem to adding fractions.

   $$4\frac{4}{10} + 1\frac{74}{100} = 4\frac{40}{100} + 1\frac{74}{100} = 5\frac{114}{100} = 6\frac{14}{100}$$

   $$1 \quad \frac{14}{100}$$

   $$4.4 + 1.74 = 6.14$$

EUREKA MATH

Lesson 13: Add decimal numbers by converting to fraction form.

235

Name _____   Date _____

1. Solve. Convert tenths to hundredths before finding the sum. Rewrite the complete number sentence in decimal form. Problems 1(a) and 1(b) are partially completed for you.

a. $5\frac{2}{10} + \frac{7}{100} = 5\frac{20}{100} + \frac{7}{100} = $ _____ $5.2 + 0.07 = $ _____	b. $5\frac{2}{10} + 3\frac{7}{100} = 8\frac{20}{100} + \frac{7}{100} = $ _____
c. $6\frac{5}{10} + \frac{1}{100}$	d. $6\frac{5}{10} + 7\frac{1}{100}$

2. Solve. Then, rewrite the complete number sentence in decimal form.

a. $4\frac{9}{10} + 5\frac{10}{100}$	b. $8\frac{7}{10} + 2\frac{65}{100}$
c. $7\frac{3}{10} + 6\frac{87}{100}$	d. $5\frac{48}{100} + 7\frac{8}{10}$

EUREKA
MATH

Lesson 13:   Add decimal numbers by converting to fraction form.

237

© 2018 Great Minds®. eureka-math.org

3. Solve by rewriting the expression in fraction form. After solving, rewrite the number sentence in decimal form.

a. $2.1 + 0.87 = 2\frac{1}{10} + \frac{87}{100}$	b. $7.2 + 2.67$
c. $7.3 + 1.8$	d. $7.3 + 1.86$
e. $6.07 + 3.93$	f. $6.87 + 3.9$
g. $8.6 + 4.67$	h. $18.62 + 14.7$

Lesson 13: Add decimal numbers by converting to fraction form.

EUREKA MATH

1. At the beginning of 2014, Jordan's height was 1.3 meters. If Jordan grew a total of 0.04 meter in 2014, what was his height at the end of the year?

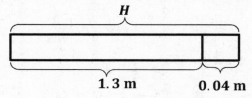

H

1.3 m   0.04 m

Jordan's height at the end of the year was 1.34 meters.

$$H = 1.3 \text{ m} + 0.04 \text{ m}$$
$$= 1\frac{30}{100}\text{ m} + \frac{4}{100}\text{ m}$$
$$= 1\frac{34}{100}\text{ m}$$
$$= 1.34 \text{ m}$$

> The tape diagram helps me to see that I need to add to solve for H, Jordan's height at the end of the year. I write the decimal numbers in fraction form using like units and then solve.

2. Tyler finished the math problem in 20.74 seconds. He beat his mom's time by 10.03 seconds. What was their combined time?

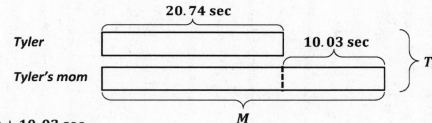

20.74 sec

Tyler

Tyler's mom   10.03 sec T

M

$$T = 20.74 \text{ sec} + 20.74 \text{ sec} + 10.03 \text{ sec}$$
$$= 20\frac{74}{100}\text{ sec} + 20\frac{74}{100}\text{ sec} + 10\frac{3}{100}\text{ sec}$$
$$= 50\frac{151}{100}\text{ sec}$$

1 sec $\frac{51}{100}$ sec

$$= 51\frac{51}{100}\text{ sec}$$
$$T = 51.51 \text{ sec}$$

Their combined time was 51.51 seconds.

Name _____   Date _____

1. The snowfall in Year 1 was 2.03 meters. The snowfall in Year 2 was 1.6 meters. How many total meters of snow fell in Years 1 and 2?

2. A deli sliced 22.6 kilograms of roast beef one week and 13.54 kilograms the next. How many total kilograms of roast beef did the deli slice in the two weeks?

Lesson 14:   Solve word problems involving the addition of measurements in decimal form.

© 2018 Great Minds®. eureka-math.org

241

3. The school cafeteria served 125.6 liters of milk on Monday and 5.34 more liters of milk on Tuesday than on Monday. How many total liters of milk were served on Monday and Tuesday?

4. Max, Maria, and Armen were a team in a relay race. Max ran his part in 17.3 seconds. Maria was 0.7 seconds slower than Max. Armen was 1.5 seconds slower than Maria. What was the total time for the team?

Lesson 14:   Solve word problems involving the addition of measurements in decimal form.

© 2018 Great Minds®. eureka-math.org

EUREKA MATH

Lesson Notes

In Grade 4, students find the sum of money amounts by expressing the amounts in unit form, adding like units (i.e., dollars + dollars and cents + cents), and then writing the answer in decimal form with a dollar sign. Writing money amounts in unit form and fraction form builds a strong conceptual foundation for decimal notation. Students are introduced to adding decimal numbers in Grade 5.

1. 4 pennies = $ __0__ . __04__   4¢ = $\frac{4}{100}$ dollar

2. 8 dimes = $ __0__ . __80__   80¢ = $\frac{8}{10}$ dollar

3. 2 quarters = $ __0__ . __50__   50¢ = $\frac{50}{100}$ dollar

> 1 penny = $\frac{1}{100}$ dollar
>
> 1 dime = $\frac{1}{10}$ dollar
>
> 1 quarter = $\frac{25}{100}$ dollar

Solve. Give the total amount of money in fraction and decimal form.

4. 7 dimes and 23 pennies

$$(7 \times 10¢) + (23 \times 1¢) = 70¢ + 23¢ = 93¢$$

$$93¢ = \frac{93}{100} \text{ dollar}$$

$$\frac{93}{100} \text{ dollar} = \$0.93$$

> 93 cents is 93 hundredths of a dollar. Thinking of that value as a fraction helps me to write it as a decimal number.

5. 1 quarter 3 dimes and 6 pennies

$$(1 \times 25¢) + (3 \times 10¢) + (6 \times 1¢) = 25¢ + 30¢ + 6¢ = 61¢$$

$$61¢ = \frac{61}{100} \text{ dollar}$$

$$\frac{61}{100} \text{ dollar} = \$0.61$$

EUREKA MATH®

6. 173 cents is what fraction of a dollar?

$\frac{173}{100}$ *dollars*

> I know that 1 cent = $\frac{1}{100}$ dollar.

Solve. Express the answer in decimal form.

> I rewrite each addend as dollars and cents. I add like units and then express the amount in decimal form.

7. 2 dollars 3 dimes 24 pennies + 3 dollars 1 quarter

2 *dollars* 54 *cents* + 3 *dollars* 25 *cents* = 5 *dollars* 79 *cents*

5 *dollars* 79 *cents* = 5$\frac{79}{100}$ *dollars* = \$5.79

8. 7 dollars 5 dimes 2 pennies + 1 dollar 3 quarters

7 *dollars* 52 *cents* + 1 *dollar* 75 *cents* = 8 *dollars* 127 *cents* = 9 *dollars* 27 *cents*

1 *dollar*   27 *cents*

9 *dollars* 27 *cents* = 9$\frac{27}{100}$ *dollars* = \$9.27

Lesson 15:   Express money amounts given in various forms as decimal numbers.

EUREKA MATH®

Name _____ Date _____

1. 100 pennies = $____._____   100¢ = ―― dollar
                                          100

2. 1 penny = $____._____   1¢ = ―― dollar
                                         100

3. 3 pennies = $____._____   3¢ = ―― dollar
                                         100

4. 20 pennies = $____._____   20¢ = ―― dollar
                                          100

5. 37 pennies = $____._____   37¢ = ―― dollar
                                          100

6. 10 dimes = $____._____   100¢ = ―― dollar
                                           10

7. 2 dimes = $____._____   20¢ = ―― dollar
                                          10

8. 4 dimes = $____._____   40¢ = ―― dollar
                                          10

9. 6 dimes = $____._____   60¢ = ―― dollar
                                          10

10. 9 dimes = $____._____   90¢ = ―― dollar
                                          10

11. 3 quarters = $____._____   75¢ = ―― dollar
                                          100

12. 2 quarters = $____._____   50¢ = ―― dollar
                                          100

13. 4 quarters = $____._____   100¢ = ―― dollar
                                           100

14. 1 quarter = $____._____   25¢ = ―― dollar
                                          100

Solve. Give the total amount of money in fraction and decimal form.

15. 5 dimes and 8 pennies

16. 3 quarters and 13 pennies

17. 3 quarters 7 dimes and 16 pennies

18. 187 cents is what fraction of a dollar?

Solve. Express the answer in decimal form.

19. 1 dollar 2 dimes 13 pennies + 2 dollars 3 quarters

20. 2 dollars 6 dimes + 2 dollars 2 quarters 16 pennies

21. 8 dollars 8 dimes + 7 dollars 1 quarter 8 dimes

Lesson 15:   Express money amounts given in various forms as decimal numbers.

EUREKA MATH

Use the RDW process to solve. Write your answer as a decimal.

1. Soo Jin needs 4 dollars 15 cents to buy a school lunch. At the bottom of her backpack, she finds 2 dollar bills, 5 quarters, and 4 pennies. How much more money does Soo Jin need to buy a school lunch?

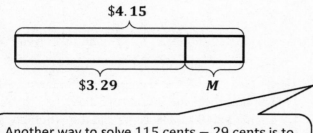

$M = 4$ *dollars* 15 *cents* $- 3$ *dollars* 29 *cents*

$= 1$ *dollar* 15 *cents* $- 29$ *cents*

           100 *cents* 15 *cents*

$= 86$ *cents*

$= \$0.86$

Soo Jin needs $\$0.86$ **more to buy a school lunch.**

Another way to solve 115 cents − 29 cents is to add 1 to each number and then solve 116 − 30. 11 tens 6 ones − 3 tens = 8 tens 6 ones.

2. Kelly has 2 quarters and 3 dimes. Jack has 5 dollars, 4 dimes, and 7 pennies. Emma has 3 dollars, 1 quarter, and 1 dime. They want to put their money together to buy a pizza that costs $11.00. Do they have enough? If not, how much more do they need?

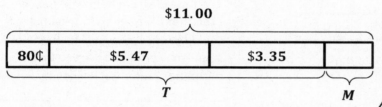

I determine how much money Kelly, Jack, and Emma each have. I add to find out how much money they have together. Then, I subtract that amount from the cost of the pizza to find out how much more money they need, M.

$T = 80$ *cents* $+ 5$ *dollars* 47 *cents* $+ 3$ *dollars* 35 *cents*

$= 8$ *dollars* 162 *cents*

              1 *dollar* 62 *cents*

$= 9$ *dollars* 62 *cents*

Kelly, Jack, and Emma have $\$9.62.$

$M = 11$ *dollars* $- 9$ *dollars* 62 *cents*

10 *dollars* 100 *cents*

$= 1$ *dollar* 38 *cents*

They do not have enough money to buy the pizza. They need $\$1.38$ **more.**

© 2018 Great Minds®. eureka-math.org

3. A pint of ice cream costs $2.49. A box of ice cream cup sundaes costs twice as much as the pint of ice cream. Brandon buys a pint of ice cream and a box of ice cream cup sundaes. How much money does he spend?

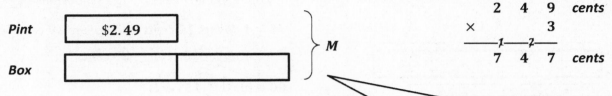

Pint $2.49

Box

M

```
    2   4   9   cents
×               3
    ─1──2─────
    7   4   7   cents
```

Brandon spends $7.47.

> I see that there are 3 units of $2.49. I rename $2.49 as 249 cents and then multiply by 3. I write my answer in decimal form.

4. Katrina has 3 dollars 28 cents. Gail has 7 dollars 52 cents. How much money does Gail need to give Katrina so that each of them has the same amount of money?

$3.28

Katrina

M

Gail

$7.52

> The tape diagram helps me to solve. I see that if Gail gives Katrina half of the difference, they will have the same amount. I subtract to find the difference, and then I divide by 2.

7 dollars 52 cents − 3 dollars 28 cents = 4 dollars 24 cents

= 424 cents

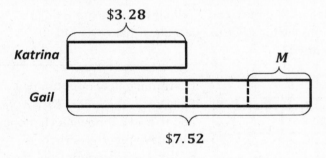

```
        2   1   2
    2 │ 4   2   4
      − 4
        0   2
          − 2
            0   4
              − 4
                0
```

212 cents = $2.12

M = $2.12

Gail needs to give Katrina $2.12 ***so that each of them has the same amount of money.***

Lesson 16:   Solve word problems involving money.

EUREKA
MATH

Name _____ Date _____

Use the RDW process to solve. Write your answer as a decimal.

1. Maria has 2 dollars, 3 dimes, and 4 pennies. Lisa has 1 dollar and 5 quarters. How much money do the two girls have in all?

2. Meiling needs 5 dollars 35 cents to buy a ticket to a show. In her wallet, she finds 2 dollar bills, 11 dimes, and 5 pennies. How much more money does Meiling need to buy the ticket?

3. Joe has 5 dimes and 4 pennies. Jamal has 2 dollars, 4 dimes, and 5 pennies. Jimmy has 6 dollars and 4 dimes. They want to put their money together to buy a book that costs $10.00. Do they have enough? If not, how much more do they need?

4. A package of mechanical pencils costs $4.99. A package of pens costs twice as much as a package of pencils. How much do a package of pens and a package of pencils cost together?

5. Carlos has 8 dollars and 48 cents. Alissa has 4 dollars and 14 cents. How much money does Carlos need to give Alissa so that each of them has the same amount of money?

Lesson 16:   Solve word problems involving money.

Grade 4
Module 7

1. Complete the tables.

a.

Yards	Feet
1	**3**
4	**12**
10	**30**

b.

Feet	Inches
1	**12**
3	**36**
9	**108**

c.

Yards	Inches
1	**36**
2	**72**
4	**144**

1 yard = 3 feet. I multiply the number of yards by 3 to find the number of feet.

1 foot = 12 inches. I multiply the number of feet by 12 to find the number of inches.

1 yard = 3 feet, and 1 foot = 12 inches. To find the number of inches in 1 yard, I can multiply, $3 \times 12 = 36$. Now I multiply the number of yards by 36 to find the number of inches.

2. Solve.

a. 3 yards 2 inches = **110** inches

There are 36 inches in 1 yard. 3×36 inches = 108 inches.

b. 12 yards 4 feet = **40** feet

There are 3 feet in 1 yard. 12×3 feet = 36 feet.

c. 3 yards 1 foot = **120** inches

I can solve this two ways: Convert yards and feet to inches, or convert yards to feet and then feet to inches.

EUREKA MATH®

Lesson 1: Create conversion tables for length, weight, and capacity units using measurement tools, and use the tables to solve problems.

253

© 2018 Great Minds®. eureka-math.org

3. Complete the table.

Pounds	Ounces
1	16
3	48
5	80

> 1 pound = 16 ounces. I multiply the number of pounds by 16 to find the number of ounces.

4. Ronald's cat weighs 9 pounds 3 ounces. How many ounces does his cat weigh?

9 *pounds* 3 *ounces*

16 oz	16 oz	16 oz	16 oz	16 oz	16 oz	16 oz	16 oz	16 oz	3 oz

T

1 *unit*: 16 *ounces*

9 *units*: 144 *ounces*

$$T = 144 \text{ ounces} + 3 \text{ ounces}$$

$$T = 147 \text{ ounces}$$

Ronald's cat weighs 147 ounces.

$$\begin{array}{r} 1\ 6 \\ \times\ \quad 9 \\ \hline 1\ 4\ 4 \end{array}$$

> I can draw a tape diagram with 9 units of 16 ounces and 1 unit of 3 ounces because the cat weighs 9 pounds 3 ounces and each pound equals 16 ounces.

> I can multiply 9 × 16 to find the number of ounces in 9 pounds. Then I can add 3 more ounces to find the total number of ounces.

5. Answer *true* or *false* for the following statement. If the statement is false, change the right side of the comparison to make it true.

2 kilograms < 1,900 grams _____*false*_____

2, 001 *grams*

> 1 kilogram = 1,000 grams
> 2 × 1,000 grams = 2,000 grams
> 2 kilograms = 2,000 grams

> The statement is false because 2,000 grams is not less than 1,900 grams. The number on the right has to be greater than 2,000.

Lesson 1: Create conversion tables for length, weight, and capacity units using measurement tools, and use the tables to solve problems.

© 2018 Great Minds®. eureka-math.org

Name _____ Date _____

1. Complete the tables.

a.

Yards	Feet
1	
2	
3	
5	
10	

b.

Feet	Inches
1	
2	
5	
10	
15	

c.

Yards	Inches
1	
3	
6	
10	
12	

2. Solve.

a.   2 yards 2 inches = _____ inches

b.   9 yards 10 inches = _____ inches

c.   4 yards 2 feet = _____ feet

d.   13 yards 1 foot = _____ feet

e.   17 feet 2 inches = _____ inches

f.   11 yards 1 foot = _____ feet

g.   15 yards 2 feet = _____ feet

h.   5 yards 2 feet = _____ inches

3. Ally has a piece of string that is 6 yards 2 feet long.   How many inches of string does she have?

Lesson 1:   Create conversion tables for length, weight, and capacity units using measurement tools, and use the tables to solve problems.

255

EUREKA MATH

4. Complete the table.

Pounds	Ounces
1	
2	
4	
10	
12	

5. Renee's baby sister weighs 7 pounds 2 ounces.   How many ounces does her sister weigh?

6. Answer *true* or *false* for the following statements.   If the statement is false, change the right side of the comparison to make it true.

a. 4 kilograms < 4,100 grams   _____

b. 10 yards < 360 inches   _____

c. 10 liters = 100,000 milliliters   _____

Create conversion tables for length, weight, and capacity units using
          measurement tools, and use the tables to solve problems.

© 2018 Great Minds®. eureka-math.org

EUREKA
MATH™

Use the RDW process to solve Problems 1 and 2.

1. Lucy buys 2 gallons of milk. How many cups of milk does she buy?

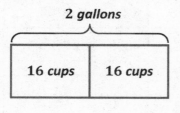

16 cups	16 cups

> I can draw a tape diagram with 2 units of 16 cups because Lucy bought 2 gallons of milk and each gallon is the same as 16 cups.

1 unit: 16 cups

2 units: 2 × 16 cups = 32 cups

Lucy has 32 cups of milk.

> I multiply 2 × 16 cups to find the number of cups in 2 gallons.

2. Matthew drank 2 liters of water today, which was 320 milliliters more water than Sarah drank today. How much water did Sarah drink today?

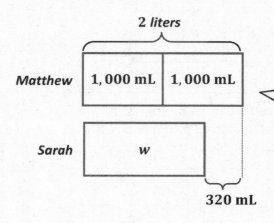

> I draw tape diagrams to represent the amount of water Matthew and Sarah drank. Matthew's tape diagram is longer than Sarah's because he drank 320 more milliliters of water than she did.

1 L = 1,000 mL

2 L = 2,000 mL

w = 2,000 mL − 320 mL

w = 1,680 mL

Sarah drank 1,680 mL of water today.

> I convert the amount of water Matthew drank, 2 liters, into milliliters. Then, I subtract from 2,000 mL the excess amount of water that Matthew drank, which is 320 mL. This tells me how much water Sarah drank.

EUREKA MATH®

Lesson 2: Create conversion tables for length, weight, and capacity units using measurement tools, and use the tables to solve problems.

257

3. Complete the tables.

a.

Gallons	Quarts
1	**4**
3	**12**
5	**20**

b.

Quarts	Pints
1	**2**
4	**8**
8	**16**

> 1 gallon = 4 quarts. I multiply the number of gallons by 4 to find the number of quarts.

> 1 quart = 2 pints. I multiply the number of quarts by 2 to find the number of pints.

4. Solve.

a. 5 gallons 3 quarts = ___**23**___ quarts

> There are 4 quarts in 1 gallon.
> 5 × 4 quarts = 20 quarts.

b. 25 gallons 2 quarts = ___**408**___ cups

> I can solve this two ways: Convert gallons and quarts to cups, or convert gallons to quarts and then all quarts to cups.

5. Answer *true* or *false* for the following statement. If your answer is false, make the statement true by correcting the right side of the comparison.

6 pints > 3 ~~quarts 1 cup~~   ___*false*___

2 *quarts* 1 *cup*

> 2 pints = 1 quart
> 3 × 2 pints = 6 pints
> 3 quarts 1 cup = 6 pints 1 cup

> The statement is false because 6 pints is not greater than 6 pints 1 cup. The number on the right has to be less than 3 quarts.

Lesson 2: Create conversion tables for length, weight, and capacity units using measurement tools, and use the tables to solve problems.

EUREKA MATH

Name _____   Date _____

Use the RDW process to solve Problems 1–3.

1. Dawn needs to pour 3 gallons of water into her fish tank. She only has a 1-cup measuring cup. How many cups of water should she put in the tank?

2. Julia has 4 gallons 2 quarts of water. Ally needs the same amount of water but only has 12 quarts. How much more water does Ally need?

3. Sean drank 2 liters of water today, which was 280 milliliters more than he drank yesterday. How much water did he drink yesterday?

4. Complete the tables.

a.

Gallons	Quarts
1	
2	
4	
12	
15	

b.

Quarts	Pints
1	
2	
6	
10	
16	

5. Solve.

a. 6 gallons 3 quarts = _____ quarts

b. 12 gallons 2 quarts = _____ quarts

c. 5 quarts 1 pint = _____ pints

d. 13 quarts 3 pints = _____ cups

e. 17 gallons 2 quarts = _____ pints

f. 27 gallons 3 quart = _____ cups

6. Explain how you solved Problem 5(f).

7. Answer true or false for the following statements. If your answer is false, make the statement true by correcting the right side of the comparison.

a. 2 quarts > 10 pints   _____

b. 6 liters = 6,000 milliliters   _____

c. 16 cups < 4 quarts 1 cup   _____

8. Joey needs to buy 3 quarts of chocolate milk. The store only sells it in pint containers. How many pints of chocolate milk should he buy? Explain how you know.

9. Granny Smith made punch. She used 2 pints of ginger ale, 3 pints of fruit punch, and 1 pint of orange juice. She served the punch in glasses that had a capacity of 1 cup. How many cups can she fill?

Lesson 2:   Create conversion tables for length, weight, and capacity units using measurement tools, and use the tables to solve problems.

© 2018 Great Minds®. eureka-math.org

EUREKA MATH

Use RDW to solve Problem 1.

1. Benjamin's football practice ends at 5:00 p.m. If practice starts at 3:00 p.m., how many minutes long is practice? Use the number line to show your work.

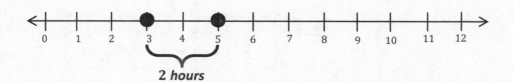

1 *hour* = 60 *minutes*

2 *hours* = 120 *minutes*

> I plot the times on the number line.
> Then, I convert the hours to minutes.

Benjamin's practice lasts for* 120 *minutes.

2. Complete the following conversion tables.

a.

Hours	Minutes
1	60
3	180
6	360

b.

Days	Hours
1	24
2	48
4	96

> 1 hour = 60 minutes
> I multiply the number of hours by 60 to find the number of minutes.

> 1 day = 24 hours
> I multiply the number of days by 24 to find the number of hours.

3. Solve

    a.   9 hours 20 minutes = **560** minutes   There are 60 minutes in 1 hour.
                                                      9×60 minutes = 540 minutes.

    b.   5 minutes 45 seconds = **345** seconds   There are 60 seconds in 1 minute.
                                                      5×60 seconds = 300 seconds.

    c.   3 days 15 hours = **87** hours   There are 24 hours in 1 day.
                                                     3×24 hours = 72 hours.

4.   In the 1860s, it took a steamship about 1 week 2 days to cross the Atlantic Ocean. How many hours are there in 1 week 2 days?

1 week 2 days

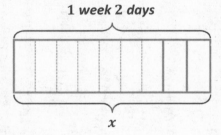

x

I can draw a tape diagram to represent 1 week 2 days. I know that there are 7 days in 1 week, so 1 week 2 days = 9 days. I can partition my tape diagram into 9 units to represent 9 days.

1 unit: **1 day = 24 hours**

9 units: 9×24 **hours = 216 hours**

$x =$ **216 hours**

$$\begin{array}{r} \overset{2}{}\overset{4}{} \\ \times \quad 9 \\ \hline 2\ 1\ 6 \end{array}$$

I can multiply 9×24 to find the total number of hours in 9 days, or 1 week 2 days.

There are 216 hours in 1 week 2 days.

    Lesson 3:   Create conversion tables for units of time, and use the tables to solve problems.

© 2018 Great Minds®. eureka-math.org

EUREKA MATH

Name _____ Date _____

Use RDW to solve Problems 1–2.

1. Jeffrey practiced his drums from 4:00 p.m. until 7:00 p.m. How many minutes did he practice? Use the number line to show your work.

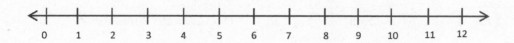

2. Isla used her computer for 5 hours over the weekend. How many minutes did she spend on the computer?

3. Complete the following conversion tables and write the rule under each table.

a.

Hours	Minutes
1	
2	
5	
9	
12	

The rule for converting hours to minutes is

_____ .

b.

Days	Hours
1	
3	
6	
8	
20	

The rule for converting days to hours is

_____ .

Lesson 3:   Create conversion tables for units of time, and use the tables to solve problems.

263

© 2018 Great Minds®. eureka-math.org

4. Solve.

   a. 10 hours 30 minutes = _____ minutes

   b. 6 minutes 15 seconds = _____ seconds

   c. 4 days 20 hours = _____ hours

   d. 3 minutes 45 seconds = _____ seconds

   e. 23 days 21 hours = _____ hours

   f. 17 hours 5 minutes = _____ minutes

5. Explain how you solved Problem 4(f).

6. It took a space shuttle 8 minutes 36 seconds to launch and reach outer space. How many seconds did it take?

7. Apollo 16's mission lasted just over 1 week 4 days. How many hours are there in 1 week 4 days?

Lesson 3:   Create conversion tables for units of time, and use the tables to solve problems.

© 2018 Great Minds®. eureka-math.org

EUREKA MATH™

Use RDW to solve the following problems.

1. Rebecca painted her bathroom in 2 hours. It took her twice as long to paint her kitchen. How many minutes did Rebecca spend painting her bathroom and kitchen?

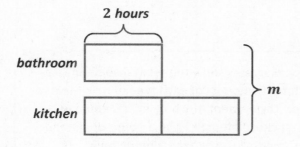

> I draw 1 unit of 2 hours to represent the amount of time Rebecca spends painting her bathroom. I draw 2 units of 2 hours to represent the amount of time she spends painting her kitchen.

1 *unit*: 2 *hours*

3 *unit*: 3 × 2 *hours* = 6 *hours*

m = 6 × 60 *minutes*

m = 360 *minutes*

Rebecca spent 360 *minutes* painting her bathroom and kitchen.

2. Mason's little sister weighed 7 pounds 9 ounces at birth. At her 6-month check-up, Mason's little sister weighed 16 pounds. How many ounces did Mason's little sister gain?

16 *pounds*

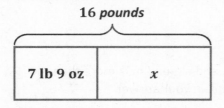

> I draw a tape diagram to represent the problem. I know a part and the whole. I subtract to find the unknown part. Then, I convert 8 pounds to ounces and add 7 more ounces.

16 *pounds* − 7 *pounds* 9 *ounces* = 8 *pounds* 7 *ounces*

15 *pounds* 16 *ounces*

x = 8 *pounds* 7 *ounces* = (8 × 16 *ounces*) + 7 *ounces* = 128 *ounces* + 7 *ounces* = 135 *ounces*

Mason's little sister gained 135 *ounces*.

Lesson 4:   Solve multiplicative comparison word problems using measurement conversion tables.

© 2018 Great Minds®. eureka-math.org

265

3. Melissa stocks 16 quarts of chocolate milk in the refrigerated case at a grocery store. She puts twice as many quarts of whole milk as chocolate milk in the case. Melissa stocks 7 fewer quarts of almond milk than whole milk in the case.

    a. How many quarts of almond milk did Melissa stock in the refrigerated case?

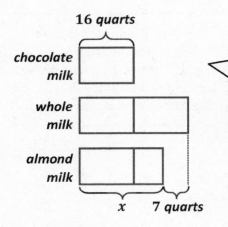

> The tape diagrams show the relationships among the different amounts of each type of milk Melissa stocked. The amount of whole milk is equal to 2 units of chocolate milk. The amount of almond milk is 7 quarts less than the almond milk.

1 unit: 16 quarts

2 units: 2 × 16 quarts = 32 quarts

x = 32 quarts − 7 quarts

x = 25 quarts

> I find the amount of whole milk by doubling the amount of chocolate milk. I find the amount of almond milk by subtracting 7 quarts from the amount of whole milk.

Melissa stocked 25 quarts of almond milk.

    b. Is the total number of quarts of chocolate milk, whole milk, and almond milk more than the 18 gallons of skim milk that are in the refrigerated case? Explain your answer.

16 quarts + 32 quarts + 25 quarts = 73 quarts

18 gallons = 18 × 4 quarts = 72 quarts

Yes, the total number of quarts of whole milk, chocolate milk, and almond milk is more than the 18 gallons of skim milk. 18 gallons is the same as 72 quarts, and the total for the other types of milk is 73 quarts. There is 1 fewer quart of skim milk than the other types of milk combined

Lesson 4:   Solve multiplicative comparison word problems using measurement
                             conversion tables.
                © 2018 Great Minds®. eureka-math.org

EUREKA MATH

Name _____   Date _____

Use RDW to solve the following problems.

1. Sandy took the train to New York City. The trip took 3 hours. Jackie took the bus, which took twice as long. How many minutes did Jackie's trip take?

2. Coleton's puppy weighed 3 pounds 8 ounces at birth. The vet weighed the puppy again at 6 months, and the puppy weighed 7 pounds. How many ounces did the puppy gain?

3. Jessie bought a 2-liter bottle of juice. Her sister drank 650 milliliters. How many milliliters were left in the bottle?

4. Hudson has a chain that is 1 yard in length. Myah's chain is 3 times as long. How many feet of chain do they have in all?

5. A box weighs 8 ounces. A shipment of boxes weighs 7 pounds. How many boxes are in the shipment?

6. Tracy's rain barrel has a capacity of 27 quarts of water. Beth's rain barrel has a capacity of twice the amount of water as Tracy's rain barrel. Trevor's rain barrel can hold 9 quarts of water less than Beth's barrel.

    a. What is the capacity of Trevor's rain barrel?

    b. If Tracy, Beth, and Trevor's rain barrels were filled to capacity, and they poured all of the water into a 30-gallon bucket, would there be enough room? Explain.

Lesson 4:   Solve multiplicative comparison word problems using measurement conversion tables.

EUREKA MATH™

Draw a tape diagram to solve the following problem.

1. Sandy bought a 3-pound bag of flour.   Adriana used 11 ounces of that flour to make cookies.   Dave used 4 ounces more of that flour than Adriana to make banana bread.   How many ounces of flour were left in Sandy's bag?

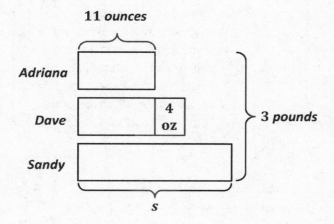

I can draw tape diagrams to represent the amount of flour Adriana and Dave used and the amount of flour Sandy still has.

11 ounces + 11 ounces + 4 ounces = 26 ounces

3 pounds = 3 × 16 ounces = 48 ounces

s = 48 ounces − 26 ounces

s = 22 ounces

Sandy has 22 ounces of flour left.

After I find the amount of flour Dave and Adriana used, which is 2 units of 11 ounces plus 4 more ounces, I convert 3 pounds to ounces and subtract.

2. Create a problem of your own using the diagram below, and solve for the unknown.

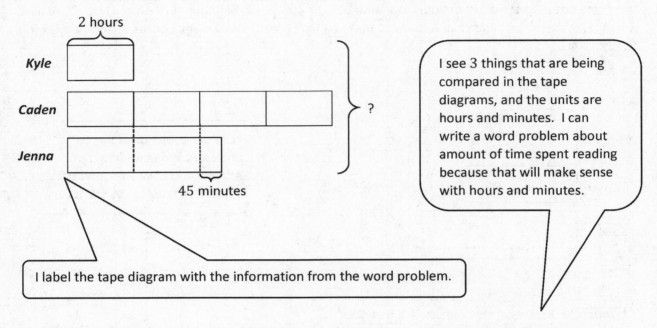

I see 3 things that are being compared in the tape diagrams, and the units are hours and minutes. I can write a word problem about amount of time spent reading because that will make sense with hours and minutes.

I label the tape diagram with the information from the word problem.

Kyle read for 2 hours last week. Caden read four times as long as Kyle read last week. Jenna read 45 minutes more than half the time that Caden read. What is the total number of minutes they read last week?

7×2 *hours* $= 14$ *hours*

14 *hours* 45 *minutes* $= (14 \times 60$ *minutes*$) + 45$ *minutes* $= 840$ *minutes* $+ 45$ *minutes* $= 885$ *minutes*

Kyle, Caden, and Jenna read for a total of 885 minutes last week.

The tape diagrams show 7 units of 2 hours plus 45 minutes, which is equal to 14 hours 45 minutes. I multiply 14×60 to convert the hours to minutes. Then, I add 45 minutes to find the total number of minutes, 885 minutes.

EUREKA MATH

Name _____   Date _____

Draw a tape diagram to solve the following problems.

1. Timmy drank 2 quarts of water yesterday. He drank twice as much water today as he drank yesterday.
    How many cups of water did Timmy drink in the two days?

2. Lisa recorded a 2-hour television show. When she watched it, she skipped the commercials. It took her
    84 minutes to watch the show. How many minutes did she save by skipping the commercials?

3. Jason bought 2 pounds of cashews. Sarah ate 9 ounces. David ate 2 ounces more than Sarah. How many
    ounces were left in Jason's bag of cashews?

4. a. Label the rest of the tape diagram below. Solve for the unknown.

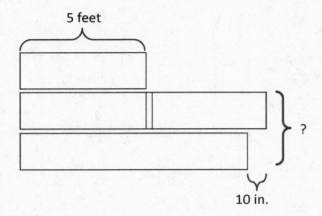

5 feet

?

10 in.

b. Write a problem of your own that could be solved using the diagram above.

---

5. Create a problem of your own using the diagram below, and solve for the unknown.

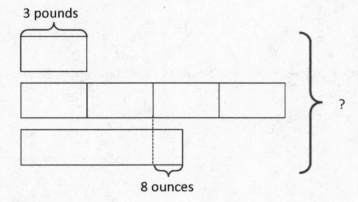

3 pounds

?

8 ounces

Lesson 5:   Share and critique peer strategies.

EUREKA MATH

> 1 gal = 8 pt
> 1 gal = 4 qt
> 1 qt = 2 pt
> 1 pt = 2 c

1. Determine the following sums and differences. Show your work.

   a. 2 gal 3 qt + 2 qt = __3__ gal __1__ qt

             1 qt 1 qt

   > I decompose and rename units to help me solve. Then, I add or subtract like units.

   b. 5 qt − 3 pt = __3__ qt __1__ pt   3 pt $\xrightarrow{+1\,pt}$ 2 qt $\xrightarrow{+3\,qt}$ 5 qt

                                   1 qt 1 pt

   > I use the arrow way counting up to 5 quarts from 3 pints. I rename 3 pints as 1 quart 1 pint and then add on 1 pint to reach 2 quarts. Finally, I add on 3 quarts to reach 5 quarts. The answer is the sum of what was added on.

   c. 7 gal 1 pt − 2 pt = __6__ gal __7__ pt

       6 gal 9 pt

   > I rename 1 gallon as 8 pints.

   d. 2 qt 3 c + 3 c = __3__ qt __2__ c   **2 qt 3 c + 3 c = 2 qt 6 c = 3 qt 2 c**

                                              1 qt 2 c

2. The capacity of the container is 4 gallons 2 quarts of liquid. Right now, 1 gallon 3 quarts of liquid are in the container. How much more liquid will the container hold?

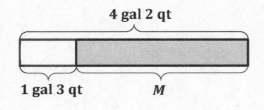

4 gal 2 qt

1 gal 3 qt   M

4 gal 2 qt − 1 gal 3 qt = 2 gal 3 qt

    3 gal 6 qt

M = 2 gal 3 qt

The container will hold 2 gallons 3 quarts more liquid.

> I rename 4 gallons 2 quarts as 3 gallons 6 quarts so that there are enough quarts to subtract 3 quarts.

EUREKA MATH®

3. Grant and Emma follow the recipe in the table to make punch.

   a. How much punch does the recipe make?

Punch Recipe	
Ingredient	**Amount**
Fruit Punch	1 gal 1 pt
Ginger Ale	2 qt 1 c
Pineapple Juice	1 gal 1 pt
Orange Sherbet	2qt

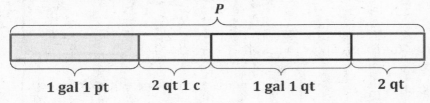

P

| 1 gal 1 pt | 2 qt 1 c | 1 gal 1 qt | 2 qt |

P = 1 gal 1 pt + 2 qt 1 c + 1 gal 1 qt + 2 qt
  = 2 gal 5 qt 1 pt 1 c

1 gal   4 c   2 c

  = 3 gal 7 c

> I could rename this as 3 gallons 1 quart 3 cups, but naming a measurement with 3 units is uncommon. I think to other measurements with 2 units: hours and minutes, weeks and days, feet and inches, pounds and ounces, and dollars and cents.

The recipe makes 3 gallons 7 cups of punch.

   b. How many more cups of liquid would they need to fill a 5-gallon container?

$$3 \text{ gal } 7 \text{ c} \xrightarrow{+9\,c} 4 \text{ gal} \xrightarrow{+16\,c} 5 \text{ gal}$$

They would need 25 more cups of liquid to fill a 5-gallon container.

> There are 16 cups in 1 gallon. I count up 9 cups to reach 4 gallons, and then I add 16 cups, or 1 gallon, to reach 5 gallons.

EUREKA MATH

Name _____   Date _____

1. Determine the following sums and differences. Show your work.

   a. 5 qt + 3 qt = _____ gal   b. 1 gal 2 qt + 2 qt = _____ gal

   c. 1 gal – 3 qt = _____ qt   d. 3 gal – 2 qt = _____ gal _____ qt

   e. 1 c + 3 c = _____ qt   f. 2 qt 3 c + 5 c = _____ qt

   g. 1 qt – 1 pt = _____ pt   h. 6 qt – 5 pt = _____ qt _____ pt

2. Find the following sums and differences. Show your work.

   a. 4 gal 2 qt + 3 qt = _____ gal _____ qt   b. 12 gal 2 qt + 5 gal 3 qt = _____ gal _____ qt

   c. 7 gal 2 pt – 3 pt = _____ gal _____ pt   d. 11 gal 3 pt – 4 gal 6 pt = _____ gal _____ pt

   e. 12 qt 5 c + 6 c = _____ qt _____ c   f. 8 gal 6 pt + 5 gal 4 pt = _____ gal _____ pt

3. The capacity of a bucket is 5 gallons. Right now, it contains 3 gallons 2 quarts of liquid. How much more liquid can the bucket hold?

4. Grace and Joyce follow the recipe in the table to make a homemade bubble solution.

   a. How much solution does the recipe make?

Homemade Bubble Solution	
Ingredient	**Amount**
Water	2 gallons 3 pints
Dish Soap	2 quarts 1 cup
Corn Syrup	2 cups

   b. How many more cups of solution would they need to fill a 4-gallon container?

Lesson 6:   Solve problems involving mixed units of capacity.

EUREKA MATH

$$1 \text{ ft} = 12 \text{ in}$$
$$1 \text{ yd} = 3 \text{ ft}$$

1. Determine the following sums and differences. Show your work.

   a.   3 yd 1 ft + 4 ft = __**4**__ yd __**2**__ ft

   3 yd 1 ft + 4 ft = 3 yd 5 ft = 4 yd 2 ft
                   ∧
              1 yd 2 ft

   > I add like units and then rename 5 feet as 1 yard 2 feet. I add 1 yard to 3 yards.

   b.   5 yd − 2 ft = __**4**__ yd __**1**__ ft
      ∧
   4 yd 3 ft

   > I rename 5 yards as 4 yards 3 feet in order to subtract 2 feet.

   c.   3 ft 7 in − 8 in = __**2**__ ft __**11**__ in
       ∧
   2 ft 19 in

   > I try to subtract like units, but I can't take 8 inches from 7 inches. I rename 3 feet 7 inches as 2 feet 19 inches by taking 1 foot from 3 feet and renaming it as 12 inches and then adding the 7 inches. Then I can subtract 8 inches.

   d.   3 ft 8 in + 4 ft 8 in = __**8**__ ft __**4**__ in

   3 ft 8 in + 4 ft 8 in = 7 ft 16 in = 8 ft 4 in
                    ∧
               1 ft 4 in

2. The height of the tree is 13 feet 8 inches. The height of the bush is 3 feet 10 inches shorter than the height of the tree. What is the height of the bush?

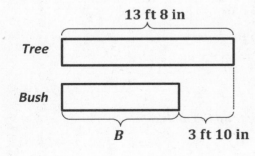

13 ft 8 in − 3 ft 10 in = 9 ft 10 in
           ∧
12 ft 20 in

B = 9 ft 10 in

The height of the bush is 9 feet 10 inches.

3. The width of Saisha's rectangular-shaped tree house is 7 feet 6 inches. The perimeter of the tree house is 35 feet.

   a. What is the length of Saisha's tree house?

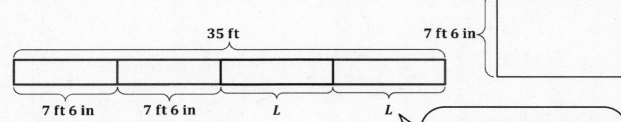

35 ft

7 ft 6 in

L

7 ft 6 in   7 ft 6 in   L   L

$7 \text{ ft } 6 \text{ in} + 7 \text{ ft } 6 \text{ in} + L + L = 35 \text{ ft}$

$14 \text{ ft } 12 \text{ in} + L + L = 35 \text{ ft}$

$15 \text{ ft} + L + L = 35 \text{ ft}$

$L + L = 20 \text{ ft}$

$L = 10 \text{ ft}$

> The tape diagram helps me to solve this problem. I see that if I subtract the widths from the perimeter that the difference is two times as much as the length.

The length of Saisha's tree house is 10 feet.

> I know the perimeter is 35 feet. I subtract the two widths from the perimeter to get the sum of the two lengths.
> $35 \text{ ft} - 15 \text{ ft} = 20 \text{ ft}$
> $10 \text{ ft} + 10 \text{ ft} = 20 \text{ ft}$

   b. How much longer is the length of Saisha's treehouse than the width?

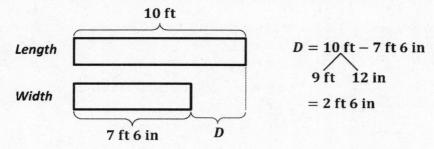

10 ft

Length

Width

7 ft 6 in   D

$D = 10 \text{ ft} - 7 \text{ ft } 6 \text{ in}$

9 ft   12 in

$= 2 \text{ ft } 6 \text{ in}$

The length of Saisha's treehouse is 2 feet 6 inches longer than the width.

Lesson 7:   Solve problems involving mixed units of length.

EUREKA MATH

Name _____   Date _____

1. Determine the following sums and differences. Show your work.

   a. 2 yd 2 ft + 1 ft = _____ yd

   b. 2 yd – 1 ft = _____ yd _____ ft

   b. 2 ft + 2 ft = _____ yd _____ ft

   d. 5 yd – 1 ft = _____ yd _____ ft

   e. 7 in + 5 in = _____ ft

   f. 7 in + 7 in = _____ ft _____ in

   g. 1 ft – 2 in = _____ in

   h. 2 ft – 6 in = _____ ft _____ in

2. Find the following sums and differences. Show your work.

   a. 4 yd 2 ft + 2 ft = _____ yd _____ ft

   b. 6 yd 2 ft + 1 yd 1 ft = _____ yd _____ ft

   c. 5 yd 1 ft – 2 ft = _____ yd _____ ft

   d. 7 yd 1 ft – 5 yd 2 ft = _____ yd _____ ft

   e. 7 ft 8 in + 5 in = _____ ft _____ in

   f. 6 ft 5 in + 5 ft 9 in = _____ ft _____ in

   g. 32 ft 3 in – 7 in = _____ ft _____ in

   h. 8 ft 2 in – 3 ft 11 in = _____ ft _____ in

3. Laurie bought 9 feet 5 inches of blue ribbon. She also bought 6 feet 4 inches of green ribbon. How much ribbon did she buy altogether?

4. The length of the room is 11 feet 6 inches. The width of the room is 2 feet 9 inches shorter than the length. What is the width of the room?

5. Tim's bedroom is 12 feet 6 inches wide. The perimeter of the rectangular-shaped bedroom is 50 feet.

    a. What is the length of Tim's bedroom?

    b. How much longer is the length of Tim's room than the width?

Lesson 7:   Solve problems involving mixed units of length.

$$1 \text{ lb} = 16 \text{ oz}$$

1. Determine the following sum and difference. Show your work.

   a. $6 \text{ lb } 7 \text{ oz} + 4 \text{ lb } 9 \text{ oz} = \underline{\hspace{0.3cm} 11 \hspace{0.3cm}} \text{ lb}$

   b. $10 \text{ lb } 4 \text{ oz} - 4 \text{ lb } 9 \text{ oz} = \underline{\hspace{0.3cm} 5 \hspace{0.3cm}} \text{ lb } \underline{\hspace{0.3cm} 11 \hspace{0.3cm}} \text{ oz}$

   6 lb 7 oz + 4 lb 9 oz = 10 lb 16 oz = 11 lb

   $$4 \text{ lb } 9 \text{ oz} \xrightarrow{+7 \text{ oz}} 5 \text{ lb} \xrightarrow{+5 \text{ lb}} 10 \text{ lb} \xrightarrow{+4 \text{ oz}} 10 \text{ lb } 4 \text{ oz}$$

   > Just like adding units of capacity or length, I add like units and rename.

   > I choose to use the arrow way to solve. I count up to reach the next whole pound. I add to find how many I count up in all. That's the same as the difference.

2. On her first birthday, Gwen weighed 23 pounds 12 ounces. On her second birthday, Gwen weighed 30 pounds 8 ounces. How much weight did Gwen gain between her first and second birthday?

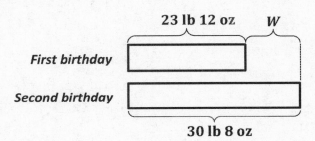

First birthday — **23 lb 12 oz** **W**

Second birthday — **30 lb 8 oz**

$$W = 30 \text{ lb } 8 \text{ oz} - 23 \text{ lb } 12 \text{ oz}$$

29 lb **24 oz**

$$= 6 \text{ lb } 12 \text{ oz}$$

> I think of 30 pounds 8 ounces as 29 pounds 16 ounces plus 8 ounces. I subtract like units to get my answer.

Gwen gained 6 pounds 12 ounces between her first and second birthday.

3. Use the information in the chart about Hayden's school supplies to answer the following question:

   On Monday, Hayden packs her supply case, a notebook, and a textbook into her empty backpack. How much does Hayden's full backpack weigh on Monday?

| Textbook 3 lb 8 oz | Supply Case 1 lb | Binder 2 lb 5 oz |
| Laptop 5 lb 12 oz | Notebook 11 oz | Backpack (empty) 2 lb 14 oz |

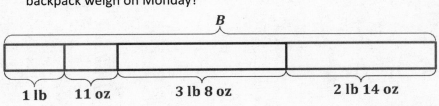

B

1 lb **11 oz** **3 lb 8 oz** **2 lb 14 oz**

$$B = 1 \text{ lb} + 11 \text{ oz} + 3 \text{ lb } 8 \text{ oz} + 2 \text{ lb } 14 \text{ oz}$$
$$= 6 \text{ lb } 33 \text{ oz}$$

2 lb **1 oz**

$$= 8 \text{ lb } 1 \text{ oz}$$

> I draw a number bond to show 33 ounces as 2 pounds 1 ounce.

Hayden's full backpack weighed 8 pounds 1 ounce on Monday.

Name _____   Date _____

1. Determine the following sums and differences. Show your work.

    a. 11 oz + 5 oz = _____ lb

    b. 1 lb 7 oz + 9 oz = _____ lb

    c. 1 lb – 11 oz = _____ oz

    d. 12 lb – 8 oz = _____ lb _____ oz

    e. 5 lb 8 oz + 9 oz = _____ lb _____ oz

    f. 21 lb 8 oz + 6 lb 9 oz = _____ lb _____ oz

    g. 23 lb 1 oz – 15 oz = _____ lb _____ oz

    h. 89 lb 2 oz – 16 lb 4 oz = _____ lb _____ oz

2. When David took his dog, Rocky, to the vet in December, Rocky weighed 29 pounds 9 ounces. When he took Rocky back to the vet in March, Rocky weighed 34 pounds 4 ounces. How much weight did Rocky gain?

3. Bianca had 6 identical jars of bubble bath. She put them all in a bag that weighed 2 ounces. The total weight of the bag filled with the six jars was 1 pound 4 ounces. How much did each jar weigh?

4. Use the information in the chart about Melissa's school supplies to answer the following questions:

   a. On Wednesdays, Melissa packs only two notebooks and a binder into her backpack. How much does her full backpack weigh on Wednesdays?

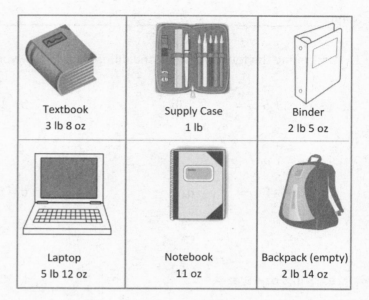

Textbook 3 lb 8 oz	Supply Case 1 lb	Binder 2 lb 5 oz
Laptop 5 lb 12 oz	Notebook 11 oz	Backpack (empty) 2 lb 14 oz

   b. On Thursdays, Melissa puts her laptop, supply case, two textbooks, and a notebook in her backpack. How much does her full backpack weigh on Thursdays?

   c. How much more does the backpack weigh with 3 textbooks and a notebook than it does with just 1 textbook and the supply case?

Lesson 8:   Solve problems involving mixed units of weight.

EUREKA MATH®

1. Determine the following sum and difference. Show your work.

    a. 6 hr 26 min + 4 hr 41 min = __11__ hr __7__ min

    6 hr 26 min + 4 hr 41 min = 10 hr 67 min = 11 hr 7 min

    1 hr   7 min

    > I add like units just as with fractions or other measurement units.

    1 day = 24 hr
    1 hr = 60 min
    1 min = 60 sec

    b. 36 min 42 sec − 24 min 56 sec = __11__ min __46__ sec

    36 min 42 sec − 24 min 56 sec = 36 min 46 sec − 25 min = 11 min 46 sec

    +4 sec   +4 sec

    > I use compensation as a strategy to solve. I add 4 seconds to each time. The difference remains the same. Subtracting just one unit, minutes, is easier than subtracting mixed units.

2. Ciera finished the race in 3 minutes 31 seconds. She beat Sarah's time by 47 seconds. What was Sarah's time?

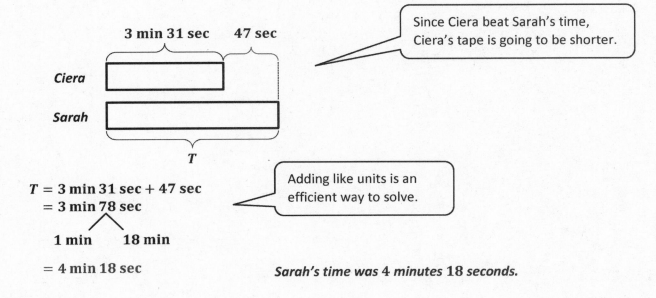

> Since Ciera beat Sarah's time, Ciera's tape is going to be shorter.

T = 3 min 31 sec + 47 sec
  = 3 min 78 sec

1 min   18 min

  = 4 min 18 sec

> Adding like units is an efficient way to solve.

Sarah's time was 4 minutes 18 seconds.

Name _____   Date _____

1. Determine the following sums and differences. Show your work.

    a. 41 min + 19 min = _____ hr

    b. 2 hr 21 min + 39 min = _____ hr

    c. 1 hr – 33 min = _____ min

    d. 3 hr – 33 min = _____ hr _____ min

    e. 31 sec + 29 sec = _____ min

    f. 5 min – 15 sec = _____ min _____ sec

2. Find the following sums and differences. Show your work.

    a. 5 hr 30 min + 35 min = _____ hr _____ min

    b. 3 hr 15 min + 5 hr 55 min = _____ hr _____ min

    c. 4 hr 4 min – 38 min = _____ hr _____ min

    d. 7 hr 3 min – 4 hr 25 min = _____ hr _____ min

    e. 3 min 20 sec + 49 sec = _____ min _____ sec

    f. 22 min 37 sec – 5 min 58 sec = _____ min _____ sec

3. It took 5 minutes 34 seconds for Melissa's oven to preheat to 350 degrees. That was 27 seconds slower than it took Ryan's oven to preheat to the same temperature. How long did it take Ryan's oven to preheat?

4. Joanna read three books. Her goal was to finish all three books in a total of 7 hours. She completed them, respectively, in 2 hours 37 minutes, 3 hours 9 minutes, and 1 hour 51 minutes.

   a. Did Joanna meet her goal? Write a statement to explain why or why not.

   b. Joanna completed the two shortest books in one evening. How long did she spend reading that evening? How long, with her goal in mind, did that leave her to read the third book?

Lesson 9: Solve problems involving mixed units of time.

EUREKA
MATH

1. On Saturday, Andrew used 1 pint 1 cup of paint from a full gallon container to paint the porch steps. On Sunday, he used twice as much paint from the container as he did on Saturday. How much paint was left in the container after Sunday?

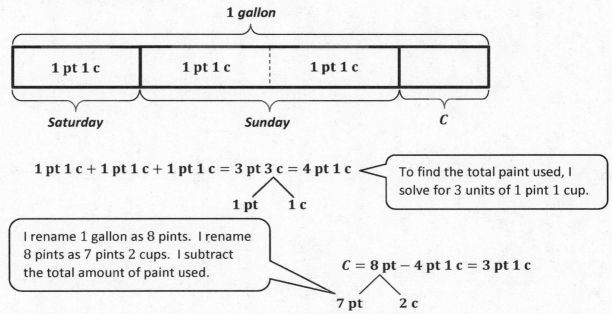

1 gallon

| 1 pt 1 c | 1 pt 1 c | 1 pt 1 c | |

Saturday *Sunday* *C*

$$1 \text{ pt } 1 \text{ c} + 1 \text{ pt } 1 \text{ c} + 1 \text{ pt } 1 \text{ c} = 3 \text{ pt } 3 \text{ c} = 4 \text{ pt } 1 \text{ c}$$

To find the total paint used, I solve for 3 units of 1 pint 1 cup.

1 pt   1 c

I rename 1 gallon as 8 pints. I rename 8 pints as 7 pints 2 cups. I subtract the total amount of paint used.

$$C = 8 \text{ pt} - 4 \text{ pt } 1 \text{ c} = 3 \text{ pt } 1 \text{ c}$$

7 pt   2 c

There were 3 pints 1 cup of paint left in the container after Sunday.

2. Shyan is 4 feet 7 inches tall. Her brother is 1 foot 5 inches taller than she is, and her sister is half as tall as her brother. How tall is Shyan's sister?

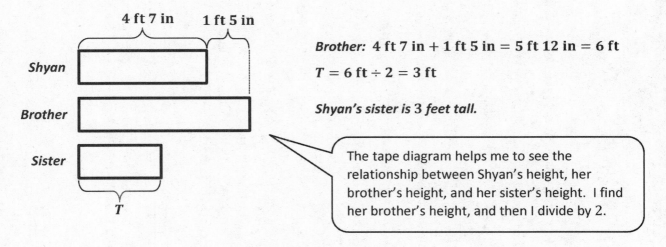

4 ft 7 in   1 ft 5 in

Shyan

Brother

Sister

T

Brother: $4 \text{ ft } 7 \text{ in} + 1 \text{ ft } 5 \text{ in} = 5 \text{ ft } 12 \text{ in} = 6 \text{ ft}$

$$T = 6 \text{ ft} \div 2 = 3 \text{ ft}$$

Shyan's sister is 3 feet tall.

The tape diagram helps me to see the relationship between Shyan's height, her brother's height, and her sister's height. I find her brother's height, and then I divide by 2.

EUREKA MATH®

Name _____ Date _____

Use RDW to solve the following problems.

1. On Saturday, Jeff used 2 quarts 1 cup of water from a full gallon to replace some water that leaked from his fish tank. On Sunday, he used 3 pints of water from the same gallon. How much water was left in the gallon after Sunday?

2. To make punch, Julia poured 1 quart 3 cups of ginger ale into a bowl and then added twice as much fruit juice. How much punch did she make in all?

3. Patti went swimming for 1 hour 15 minutes on Monday. On Tuesday, she swam twice as long as she swam on Monday. On Wednesday, she swam 50 minutes less than the time she swam on Tuesday. How much time did she spend swimming during that three-day period?

4. Myah is 4 feet 2 inches tall. Her sister, Ally, is 10 inches taller. Their little brother is half as tall as Ally. How tall is their little brother in feet and inches?

5. Rick and Laurie have three dogs. Diesel weighs 89 pounds 12 ounces. Ebony weighs 33 pounds 14 ounces less than Diesel. Luna is the smallest at 10 pounds 2 ounces. What is the combined weight of the three dogs in pounds and ounces?

1. A rectangular sidewalk is 2 feet 9 inches wide. Its length is three times the width plus 5 more inches. How long is the sidewalk?

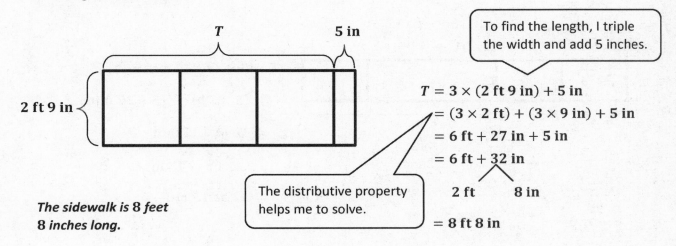

To find the length, I triple the width and add 5 inches.

$T = 3 \times (2 \text{ ft } 9 \text{ in}) + 5 \text{ in}$
$= (3 \times 2 \text{ ft}) + (3 \times 9 \text{ in}) + 5 \text{ in}$
$= 6 \text{ ft} + 27 \text{ in} + 5 \text{ in}$
$= 6 \text{ ft} + 32 \text{ in}$

2 ft   8 in

The distributive property helps me to solve.

$= 8 \text{ ft } 8 \text{ in}$

The sidewalk is 8 feet 8 inches long.

2. Mr. Lalonde plans to make his world-famous cookies. He has 2 pounds 3 ounces of brown sugar. This is $\frac{1}{3}$ of the total amount of brown sugar needed. If he uses 7 ounces of brown sugar for each batch of cookies, how many batches of cookies can he make?

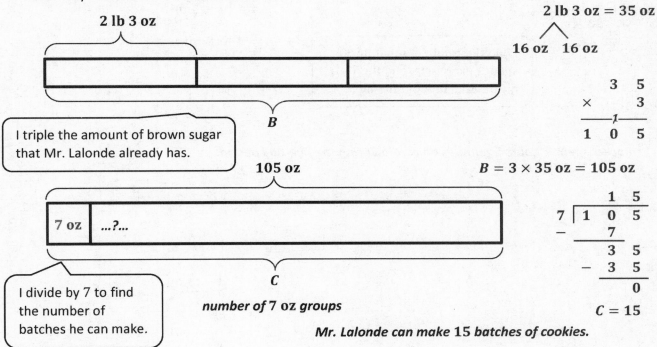

2 lb 3 oz

I triple the amount of brown sugar that Mr. Lalonde already has.

2 lb 3 oz = 35 oz

16 oz   16 oz

$\begin{array}{r} 3\ 5 \\ \times\quad 3 \\ \hline 1\ 0\ 5 \end{array}$

$B = 3 \times 35 \text{ oz} = 105 \text{ oz}$

105 oz

7 oz   …?…

I divide by 7 to find the number of batches he can make.

number of 7 oz groups

$\begin{array}{r} 1\ 5 \\ 7\overline{)1\ 0\ 5} \\ -\ 7 \\ \hline 3\ 5 \\ -\ 3\ 5 \\ \hline 0 \end{array}$

$C = 15$

Mr. Lalonde can make 15 batches of cookies.

EUREKA MATH®

3. Rocket exercised for 2 hours 27 minutes each day for 5 days. He spent an equal amount of time on lower body, upper body, and cardio. How long did he spend on cardio during the five-day period?

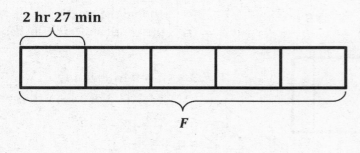

2 hr 27 min

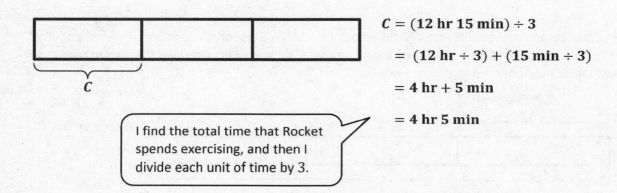

$$F = 5 \times 2 \text{ hr } 27 \text{ min}$$
$$= (5 \times 2 \text{ hr}) + (5 \times 27 \text{ min})$$
$$= 10 \text{ hr } 135 \text{ min}$$

            2 hr   **15 min**

$$= 12 \text{ hr } 15 \text{ min}$$

C

$$C = (12 \text{ hr } 15 \text{ min}) \div 3$$
$$= (12 \text{ hr} \div 3) + (15 \text{ min} \div 3)$$
$$= 4 \text{ hr} + 5 \text{ min}$$
$$= 4 \text{ hr } 5 \text{ min}$$

I find the total time that Rocket spends exercising, and then I divide each unit of time by 3.

Rocket spent 4 hours 5 minutes on cardio during the five-day period.

     Lesson 11:   Solve multi-step measurement word problems.

EUREKA MATH

Name _____   Date _____

Use RDW to solve the following problems.

1. Ashley ran a marathon and finished 1 hour 40 minutes after P.J., who had a time of 2 hours 15 minutes. Kerry finished 12 minutes before Ashley. How long did it take Kerry to run the marathon?

2. Mr. Foote's deck is 12 ft 6 in wide. Its length is twice the width plus 3 more inches. How long is the deck?

3. Mrs. Lorentz bought 12 pounds 8 ounces of sugar. This is $\frac{1}{4}$ of the sugar she will use to make sugar cookies in her bakery this week. If she uses 10 ounces of sugar for each batch of sugar cookies, how many batches of sugar cookies will she make in a week?

4. Beth Ann practiced piano for 1 hour 5 minutes each day for 1 week. She had 5 songs to practice and spent the same amount of time practicing each song. How long did she practice each song during the week?

5. The concession stand has 18 gallons of punch. If there are a total of 240 students who want to purchase 1 cup of punch each, will there be enough punch for everyone?

1. Draw a tape diagram to show $1\frac{2}{3}$ yards = 5 feet.

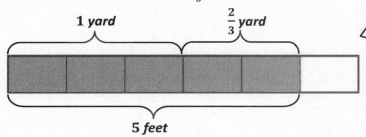

1 yard $\frac{2}{3}$ **yard**

5 feet

> I know that 1 yard = 3 feet, so I can decompose each yard in my tape diagram into 3 feet. I can shade in $1\frac{2}{3}$ yards, and since each unit is $\frac{1}{3}$ yard, or 1 foot, I can see that $1\frac{2}{3}$ yards is equal to 5 feet.

2. Solve the problems using whatever tool works best for you.

1 foot

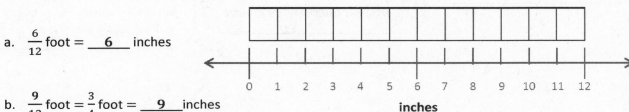

0 1 2 3 4 5 6 7 8 9 10 11 12

inches

a. $\frac{6}{12}$ foot = ___6___ inches

b. $\frac{9}{12}$ foot = $\frac{3}{4}$ foot = ___9___ inches

c. $\frac{8}{12}$ foot = $\frac{4}{6}$ foot = ___8___ inches

> For part (a), I know that $\frac{6}{12}$ foot = $\frac{1}{2}$ foot, and I know that half a foot is 6 inches. For parts (b) and (c), I can make equivalent fractions and then find the number of inches. $\frac{3 \times 3}{4 \times 3} = \frac{9}{12}$. $\frac{9}{12}$ foot is the same as 9 inches.

3. Solve.

a. $5\frac{1}{3}$ yd = ___**16**___ ft	b. $4\frac{3}{4}$ gal = ___**19**___ qt	c. $3\frac{1}{3}$ ft = ___**40**___ in
15 **1**	**16** **3**	**36** **4**
feet *foot*	*quarts* *quarts*	*inches* *inches*

1 yard = 3 feet, so 5 yards = 5×3 feet = 15 feet. And $\frac{1}{3}$ yard = 1 foot. 15 feet + 1 foot = 16 feet.

1 gallon = 4 quarts, so 4 gallons = 4×4 quarts = 16 quarts. And $\frac{1}{4}$ gallon = 1 quart, so $\frac{3}{4}$ gallon = 3 quarts. 16 quarts + 3 quarts = 19 quarts.

1 foot = 12 inches, so 3 feet = 3×12 inches = 36 inches. And $\frac{1}{12}$ foot = 1 inch, so $\frac{1}{3} = \frac{1 \times 4}{3 \times 4} = \frac{4}{12}$. $\frac{4}{12}$ foot equals 4 inches. 36 inches + 4 inches = 40 inches.

Lesson 12:   Use measurement tools to convert mixed number measurements to smaller units.

EUREKA MATH

Name _____ Date _____

1. Draw a tape diagram to show $1\frac{1}{3}$ yards = 4 feet.

2. Draw a tape diagram to show $\frac{1}{2}$ gallon = 2 quarts.

3. Draw a tape diagram to show $1\frac{3}{4}$ gallons = 7 quarts.

4. Solve the problems using whatever tool works best for you.

   a. $\frac{1}{2}$ foot= _____ inches

   b. $\frac{1}{12}$ foot = $\frac{1}{4}$ foot = _____ inches

   c. $\frac{1}{12}$ foot = $\frac{1}{6}$ foot = _____ inches

   d. $\frac{1}{12}$ foot = $\frac{1}{3}$ foot= _____ inches

   e. $\frac{1}{12}$ foot = $\frac{2}{3}$ foot = _____ inches

   f. $\frac{1}{12}$ foot = $\frac{5}{6}$ foot = _____ inches

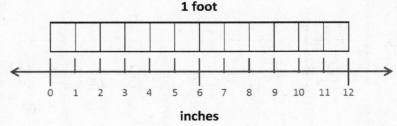

1 foot

inches

EUREKA MATH

Lesson 12: Use measurement tools to convert mixed number measurements to smaller units.

© 2018 Great Minds®. eureka-math.org

299

5. Solve.

a. $2\frac{2}{3}$ yd = _____ ft	b. $3\frac{1}{3}$ yd = _____ pie
c. $3\frac{1}{2}$ gal = _____ qt	d. $5\frac{1}{4}$ gal = _____ qt
e. $6\frac{1}{4}$ ft = _____ in	f. $7\frac{1}{3}$ ft = _____ in
g. $2\frac{1}{2}$ ft = _____ in	h. $5\frac{3}{4}$ ft = _____ in
i. $9\frac{2}{3}$ ft = _____ in	j. $7\frac{5}{6}$ ft = _____ in

Lesson 12:   Use measurement tools to convert mixed number measurements to smaller units.

© 2018 Great Minds®. eureka-math.org

EUREKA MATH®

1. Solve.

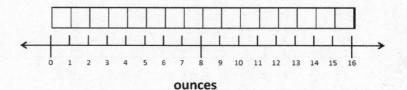

ounces

a. $\frac{2}{16}$ pound = ____2____ ounces

b. $\frac{8}{16}$ pound = $\frac{2}{4}$ pound = ____8____ ounces

c. $\frac{6}{16}$ pound = $\frac{3}{8}$ pound = ____6____ ounces

> For part (a), I know that $\frac{1}{16}$ pound = 1 ounce, so $\frac{2}{16}$ pound = 2 ounces. For part (b), I know that $\frac{2}{4}$ pound = $\frac{1}{2}$ pound, which is equal to $\frac{8}{16}$ pound or 8 ounces. For part (c), I can make equivalent fractions. $\frac{3 \times 2}{8 \times 2} = \frac{6}{16}$. And $\frac{6}{16}$ pound = 6 ounces.

2. Draw a tape diagram to show $1\frac{1}{8}$ pounds = 18 ounces

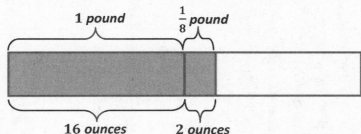

16 *ounces* + 2 *ounces* = 18 *ounces*

> I can draw a tape diagram that shows $1\frac{1}{8}$ pounds. Then I can convert the pounds to ounces. 1 pound = 16 ounces. I can use an equivalent fraction to figure out how many ounces are in $\frac{1}{8}$ pound. $\frac{1 \times 2}{8 \times 2} = \frac{2}{16}$, so $\frac{1}{8}$ pound = 2 ounces.

3. Solve.

1 hour

0 1 2 3 4 5 6 7 8 9 10 11 12 13 14 15 16 17 18 19 20 21 22 23 24 25 26 27 28 29 30 31 32 33 34 35 36 37 38 39 40 41 42 43 44 45 46 47 48 49 50 51 52 53 54 55 56 57 58 59 60

minutes

a. $\frac{45}{60}$ hour $= \frac{3}{4}$ hour $=$ ___**45**___ minutes

b. $\frac{40}{60}$ hour $= \frac{2}{3}$ hour $=$ ___**40**___ minutes

> For part (a), I know that $\frac{1}{4}$ hour = 15 minutes, so $\frac{3}{4}$ hour = 45 minutes = $\frac{45}{60}$ hour.
>
> For part (b), I know that $\frac{1}{3}$ hour = 20 minutes, so $\frac{2}{3}$ hour = 40 minutes = $\frac{40}{60}$ hour.

4. Solve.

a. $3\frac{5}{8}$ pounds $=$ ___**58**___ ounces

48 oz   **10 oz**

b. $4\frac{1}{4}$ lb $=$ ___**68**___ oz

64 oz   **4 oz**

c. $2\frac{3}{4}$ hours $=$ ___**165**___ minutes

120 min   **45 min**

> 1 pound = 16 ounces, so 3 pounds = 3 × 16 ounces = 48 ounces. And $\frac{1}{8}$ pound = 2 ounces, so $\frac{5}{8}$ pound = 10 ounces. 48 ounces + 10 ounces = 58 ounces.

> 4 pounds = 4 × 16 ounces = 64 ounces. And $\frac{1}{4}$ pound = 4 ounces. 64 ounces + 4 ounces = 68 ounces.

> 1 hour = 60 minutes, so 2 hours = 2 × 60 minutes = 120 minutes. And $\frac{1}{4}$ hour = 15 minutes, so $\frac{3}{4}$ hour = 45 minutes. 120 minutes + 45 minutes = 165 minutes.

Lesson 13: Use measurement tools to convert mixed number measurements to smaller units.

EUREKA MATH®

Name _____ Date _____

1. Solve.

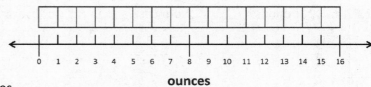

   a. $\frac{1}{16}$ pound = _____ ounce

   b. $\frac{}{16}$ pound = $\frac{1}{2}$ pound = _____ ounces

   c. $\frac{}{16}$ pound = $\frac{1}{4}$ pound = _____ ounces

   d. $\frac{}{16}$ pound = $\frac{3}{4}$ pound = _____ ounces

   e. $\frac{}{16}$ pound = $\frac{1}{8}$ pound = _____ ounces

   f. $\frac{}{16}$ pound = $\frac{5}{8}$ pound = _____ ounces

2. Draw a tape diagram to show $1\frac{1}{4}$ pounds = 20 ounces.

3. Solve.

1 hour

minutes

   a. $\frac{1}{60}$ hour = _____ minute

   b. $\frac{}{60}$ hour = $\frac{1}{2}$ hour = _____ minutes

   c. $\frac{}{60}$ hour = $\frac{1}{4}$ hour = _____ minutes

   d. $\frac{}{60}$ hour = $\frac{1}{3}$ hour = _____ minutes

4. Draw a tape diagram to show that $2\frac{1}{4}$ hours = 135 minutes.

EUREKA MATH®

Lesson 13:   Use measurement tools to convert mixed number measurements to smaller units.

© 2018 Great Minds®. eureka-math.org

303

5. Solve.

a. $2\frac{1}{4}$ pounds = _____ ounces	b. $4\frac{7}{8}$ pounds = _____ ounces
c. $6\frac{3}{4}$ lb = _____ oz	d. $4\frac{1}{8}$ lb = _____ oz
e. $1\frac{3}{4}$ hours = _____ minutes	f. $4\frac{1}{2}$ hours = _____ minutes
g. $3\frac{3}{4}$ hr = _____ min	h. $5\frac{1}{3}$ hr = _____ min
i. $4\frac{2}{3}$ yards = _____ feet	j. $6\frac{1}{3}$ yd = _____ ft
k. $4\frac{1}{4}$ gallons = _____ quarts	l. $2\frac{3}{4}$ gal = _____ qt
m. $6\frac{1}{4}$ feet = _____ inches	n. $9\frac{5}{6}$ ft = _____ in

© 2018 Great Minds®. eureka-math.org

EUREKA
MATH®

Use RDW to solve the following problems.

1. Doug practiced piano for 1 hour and 50 minutes on Monday. On Tuesday, he practiced piano for 25 minutes less than Monday. How many minutes did Doug practice piano on Monday and Tuesday?

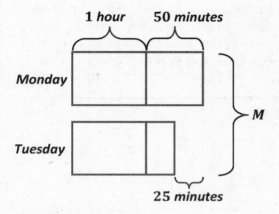

I can draw a tape diagram to represent the amount of time that Doug practiced piano each day. The tape for Monday is longer than Tuesday's because he practiced for 25 minutes less on Tuesday.

1 *hour* 50 *minutes* − 25 *minutes* = 1 *hour* 25 *minutes*

I subtract 25 minutes from Monday's time to figure out how long Doug practiced on Tuesday.

1 *hour* 50 *minutes* + 1 *hour* 25 *minutes* = 2 *hours* 75 *minutes*

2 *hours* 75 *minutes* = 120 *minutes* + 75 *minutes* = 195 *minutes*

M = 195 *minutes*

I add the times for Monday and Tuesday to find the total time. Then I convert the hours to minutes. 1 hour = 60 minutes, so 2 hours = 120 minutes.

Doug practiced piano for 195 minutes on Monday and Tuesday.

EUREKA MATH®   Lesson 14:   Solve multi-step word problems involving converting mixed number measurements to a single unit.   305

© 2018 Great Minds®. eureka-math.org

2. Ella can make 15 bracelets from a 105-inch piece of cord.

   a. How many inches of cord would be needed to make 60 bracelets?

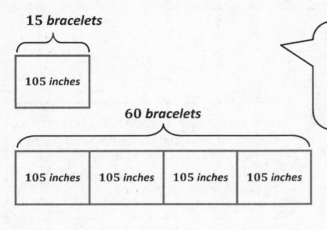

15 bracelets

105 *inches*

60 bracelets

| 105 *inches* | 105 *inches* | 105 *inches* | 105 *inches* |

> I can draw 1 unit of 105 inches to represent the length of cord needed to make 15 bracelets. I can draw 4 units of 105 inches to represent the length of cord needed to make 60 bracelets because $4 \times 15 = 60$.

4×105 *inches* $= 420$ *inches*

$$
\begin{array}{r}
1\ 0\ 5 \\
\times \qquad 4 \\
\hline
4\ 2\ 0
\end{array}
$$

> To find the total number of inches of cord that Ella needs to make 60 bracelets, I can multiply 4×105 inches.

Ella needs 420 *inches of cord to make* 60 *bracelets.*

   b. Extension: The cord Ella uses to make bracelets is also sold in $8\frac{1}{3}$-foot packages. How many packages would be needed to make 60 bracelets?

$8\frac{1}{3}$ *feet* $= 100$ *inches*

96   4
inches   *inches*

> I can covert $8\frac{1}{3}$ feet to inches. 8×12 inches $= 96$ inches and $\frac{1}{3}$ foot $= 4$ inches. 96 inches $+ 4$ inches $= 100$ inches. Ella would need to buy 5 packages because 4 packages would only be 400 inches of cord and she needs 420 inches of cord.

5×100 *inches* $= 500$ *inches*

Ella would need 5 *packages to make* 60 *bracelets.*

     Lesson 14:   Solve multi-step word problems involving converting mixed number measurements to a single unit.

© 2018 Great Minds®. eureka-math.org

EUREKA MATH

Name _____   Date _____

Use RDW to solve the following problems.

1.   Molly baked a pie for 1 hour and 45 minutes. Then, she baked banana bread for 35 minutes less than the pie. How many minutes did it take to bake the pie and the bread?

2.   A slide on the playground is $12\frac{1}{2}$ feet long. It is 3 feet 7 inches longer than the small slide. How long is the small slide?

3.   The fish tank holds 8 gallons 2 quarts of water. Jeffrey poured $1\frac{3}{4}$ gallons into the empty tank. How much more water does he still need to pour into the tank to fill it?

Lesson 14:   Solve multi-step word problems involving converting mixed number measurements to a single unit.

© 2018 Great Minds®. eureka-math.org

307

4. The candy shop puts 10 ounces of gummy bears in each box. How many boxes do they need to fill if there are $21\frac{1}{4}$ pounds of gummy bears?

5. Mom can make 10 brownies from a 12-ounce package.

    a. How many ounces of brownie mix would be needed to make 50 brownies?

    b. Extension: The brownie mix is also sold in $1\frac{1}{2}$ pound bags. How many bags would be needed to make 120 brownies?

Lesson 14: Solve multi-step word problems involving converting mixed number measurements to a single unit.

EUREKA MATH

1. Find the area of the figure that is shaded.

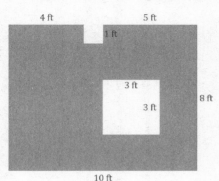

> I find the area of the white portion inside the shaded figure and the area of the cutout.

3 ft × 3 ft = 9 square ft

1 ft × 1 ft = 1 square ft

9 square ft + 1 square ft = 10 square ft

10 ft × 8 ft = 80 square ft

> I think of the shaded area as a rectangle without the cutouts and find its area.

80 square ft − 10 square ft = 70 square ft

> I subtract the area of the cutouts from the area of the larger rectangle to find the area of the figure that is shaded.

The area of the shaded figure is **70** *square feet.*

2. A wall is 10 feet tall and 12 feet wide. A window with a width of 2 feet and a height of 4 feet is in the center of the wall. Find the area of the wall that can be painted.

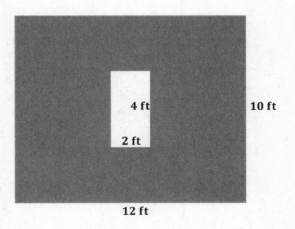

12 ft × 10 ft = 120 square ft

2 ft × 4 ft = 8 square ft

120 square ft − 8 square ft = 112 square ft

The area of the wall that can be painted is **112** *square feet.*

EUREKA MATH

Lesson 15:   Create and determine the area of composite figures.

309

© 2018 Great Minds®. eureka-math.org

Name _____   Date _____

For homework, complete the top portion of each page. This will become an answer key for you to refer to when completing the bottom portion as a mini-personal white board activity during the summer.

Find the area of the figure that is shaded.

1.

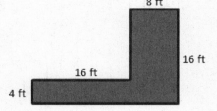

2.

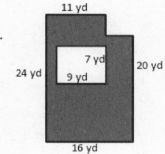

- -

Find the area of the figure that is shaded.

1.

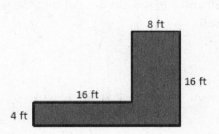

2.

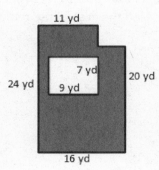

Challenge: Replace the given dimensions with different measurements, and solve again.

3. A wall is 8 feet tall and 19 feet wide. An opening 7 feet tall and 8 feet wide was cut into the wall for a doorway. Find the area of the remaining portion of the wall.

- -

3. A wall is 8 feet tall and 19 feet wide. An opening 7 feet tall and 8 feet wide was cut into the wall for a doorway. Find the area of the remaining portion of the wall.

Lesson 15:   Create and determine the area of composite figures.

EUREKA
MATH®

1. Use a ruler and protractor to create and shade a figure according to the directions:

   Draw a rectangle that is 15 centimeters long and 5 centimeters wide. Inside the rectangle, draw a smaller rectangle that is 10 centimeters long and 4 centimeters wide. Inside the smaller rectangle, draw a square that has side lengths of 2 centimeters. Shade the larger rectangle and the square.

   Find the area of the shaded space.

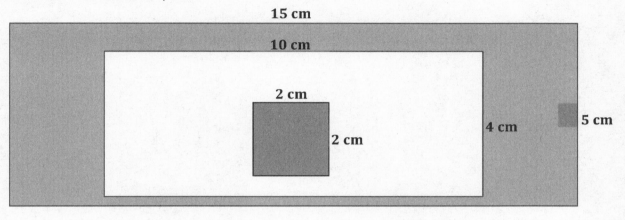

> To find the area of the shaded space, I subtract the area of the smaller, unshaded rectangle from the area of the larger, shaded rectangle, and add back the area of the square.

Large rectangle: $15 \text{ cm} \times 10 \text{ cm} = 150$ **square cm**

Small rectangle: $10 \text{ cm} \times 4 \text{ cm} = 40$ **square cm**

 150 **square cm** $- 40$ **square cm** $= 110$ **square cm**

Square: $2 \text{ cm} \times 2 \text{ cm} = 4$ **square cm**

 110 **square cm** $+ 4$ **square cm** $= 114$ **square cm**

The area of the shaded space is 114 *square centimeters.*

2. Zachary hangs a television that is 4 feet long and 2 feet wide on a wall that is 10 feet long and 8 feet tall. How much area of the wall is not covered up by the television?

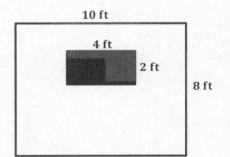

Wall: $8 \text{ ft} \times 10 \text{ ft} = 80$ **square ft**

TV: $2 \text{ ft} \times 4 \text{ ft} = 8$ **square ft**

 80 **square ft** $- 8$ **square ft** $= 72$ **square ft**

 72 *square feet of the wall is not covered by the television.*

Name _____   Date _____

For homework, complete the top portion of each page. This will become an answer key for you to refer to when completing the bottom portion as a mini-personal white board activity during the summer.

Use a ruler and protractor to create and shade a figure according to the directions. Then, find the area of the unshaded part of the figure.

1. Draw a rectangle that is 18 cm long and 6 cm wide. Inside the rectangle, draw a smaller rectangle that is 8 cm long and 4 cm wide. Inside the smaller rectangle, draw a square that has a side length of 3 cm. Shade in the smaller rectangle, but leave the square unshaded. Find the area of the unshaded space.

- -

1. Draw a rectangle that is 18 cm long and 6 cm wide. Inside the rectangle, draw a smaller rectangle that is 8 cm long and 4 cm wide. Inside the smaller rectangle, draw a square that has a side length of 3 cm. Shade in the smaller rectangle, but leave the square unshaded. Find the area of the unshaded space.

Lesson 16:   Create and determine the area of composite figures.

© 2018 Great Minds®. eureka-math.org

315

2. Emanuel's science project display board is 42 inches long and 48 inches wide. He put a 6-inch border around the edge inside the board and placed a title in the center of the board that is 22 inches long and 6 inches wide. How many square inches of open space does Emanuel have left on his board?

- -

2. Emanuel's science project display board is 42 inches long and 48 inches wide. He put a 6-inch border around the edge inside the board and placed a title in the center of the board that is 22 inches long and 6 inches wide. How many square inches of open space does Emanuel have left on his board?

Challenge: Replace the given dimensions with different measurements, and solve again.

Lesson 16:   Create and determine the area of composite figures.

EUREKA MATH®

1. Plot and label each point on the number line below, and complete the chart.

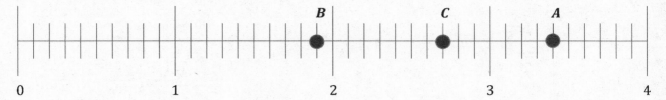

Point	Unit Form	Decimal Form	Mixed Number (ones and fraction form)	How much more to get to the next whole number?
A	3 ones 4 tenths	3.4	$3\frac{4}{10}$	0.6
B	**1 one 9 tenths**	1.9	$1\frac{9}{10}$	0.1
C	**2 ones 7 tenths**	2.7	$2\frac{7}{10}$	$\frac{3}{10}$ or 0.3

> To solve for point C, I named two and seven tenths, but I could have named any decimal that is 3 tenths from a whole number between zero and four: 0.7, 1.7, or 3.7.

2. Complete the chart.

Decimal	Mixed Number	Tenths	Hundredths
5.8	$5\frac{8}{10}$	58 *tenths or* $\frac{58}{10}$	580 *hundredths or* $\frac{580}{100}$
9.2	$9\frac{2}{10}$	92 *tenths or* $\frac{92}{10}$	920 *hundredths or* $\frac{920}{100}$

> I convert 9.2 to $9\frac{20}{100}$ to help me write the number as hundredths.

Name _____   Date _____

1. Decimal Fraction Review: Plot and label each point on the number line below, and complete the chart.
    Only solve the portion above the dotted line.

Point	Unit Form	Decimal Form	Mixed Number (ones and fraction form)	How much more to get to the next whole number?
A	2 ones and 9 tenths			
B		4.4	$4\frac{4}{10}$	
C				$\frac{2}{10}$ or 0.2

- -

1. Complete the chart. Create your own problem for B, and plot the point.

Point	Unit Form	Decimal Form	Mixed Number (ones and fraction form)	How much more to get to the next whole number?
A	2 ones and 9 tenths			
B				

2. Complete the chart. The first one has been done for you. Only solve the top portion above the dotted line.

Decimal	Mixed Number	Tenths	Hundredths
3.2	$3\frac{2}{10}$	32 tenths or $\frac{32}{10}$	320 hundredths or $\frac{320}{100}$
8.6			
11.7			
4.8			

- -

2. Complete the chart. Create your own problem in the last row.

Decimal	Mixed Number	Tenths	Hundredths
3.2			
8.6			
11.7			

Lesson 17:   Practice and solidify Grade 4 fluency.

EUREKA MATH

Credits

Great Minds® has made every effort to obtain permission for the reprinting of all copyrighted material. If any owner of copyrighted material is not acknowledged herein, please contact Great Minds for proper acknowledgment in all future editions and reprints of this module.